Quantitative Modeling of
Soil Forming Processes

Related Society Publications

Advances in Measurement of Soil Physical Properties: Bringing Theory into Practice

Factors of Soil Formation: A Fiftieth Anniversary Retrospective

Quantitative Methods in Soil Mineralogy

Rates of Soil Chemical Processes

Spatial Variabilities of Soils and Landforms

For information on these titles, please contact the ASA, CSSA, SSSA Headquarters Office; Attn: Marketing; 677 South Segoe Road; Madison, WI 53711–1086. Phone: (608) 273–8080. Fax: (608) 273–2021.

Quantitative Modeling of Soil Forming Processes

Proceedings of a symposium sponsored by Divisions S-5 and S-9 of the Soil Science Society of America in Minneapolis, Minnesota, 2 Nov. 1992.

Editors
R.B. Bryant and R.W. Arnold

Organizing Committee
R.B. Bryant
R.W. Arnold
M.R. Hoosbeek

Editor-in Chief SSSA
J.M. Bigham

Managing Editor
Jon M. Bartels

SSSA Special Publication Number **39**

Soil Science Society of America, Inc.

Madison, Wisconsin, USA

1994

Cover design is adapted from a 1992 article by M.R. Hoosbeek and R.B. Bryant in *Geoderma* (55:183–210).

Soil Science Society of America, Inc.
677 South Segoe Road, Madison, WI 53711, USA

Library of Congress Cataloging-in-Publication Data
Quantitative modeling of soil forming processes : proceedings of a
    symposium sponsored by Divisions S-5 and S-9 of the Soil Science
    Society of America in Minneapolis, Minnesota, 2 Nov. 1992 / editors.
    R.B. Bryant and R.W. Arnold : organizing committee, R.B. Bryant,
    R.W. Arnold, M.R. Hoosbeek : editor-in-chief, SSSA, J.M. Bigham.
        p.   cm.—(SSSA special publication : no. 39)
    Includes bibliographical references.
    ISBN 0-89118-814-2
    1. Soil Formation—Simulation methods—Congresses.   I. Bryant,
Ray Baldwin, 1951-   . II. Arnold, R. W. (Richard W.) III. Soil
Science Society of America. Division S-5.   IV. Soil Science Society
of America. Division S-9.   V. Series.
    S592.5.Q36   1994
    551.3'05'011—dc20   94-31647
                                                                   CIP

Printed in the United States of America

CONTENTS

FOREWORD

Spatial variability with accompanying mechanisms, sources and dynamics are the life blood of geoscientists. Such information is prerequisite for extrapolation beyond site-specific locale. For over 100 years pedologists have been productively engaged in defining parameters that govern time and spatial distributions of soil–landform patterns. Over the past 50 years there has been a strong quest for quantification of these soil–landform relationships. Numerous conceptual, verbal, structural and mathematical models have been developed with variable success in generalizing the complexities of soil within the pedosphere. This continues to be the challenge of the future. Models by their very nature are simplistic, mental constructs of reality. They are generalized and commonly imprecise. If not allowed to self-destruct in part or in total, the model may be self-limiting and retard rather than enhance our perception and knowledge of reality.

This publication accurately documents the new generation of evolving models to simulate soil formation processes, rate dynamics and patterns of spatial diversity. Ideally such simulation models should enhance our ability to scale observations from variable levels of resolution, target information voids, capture intuitive knowledge in a more rigorous analytical framework, and synergistically integrate team efforts among field and modeling expertise. These are ambitious goals, but hopefully the spiral of progress is upward. This document further illustrates the value of coupling theory and practice in physical and biological application of simulation models. Likewise, it illustrates how sophisticated statistical and Geographic Information Systems (GIS) tools may be used to quantify the complex interconnective spatial and temporal dynamics of soil systems. Certainly in this era of new mandates for stewardship of the biosphere, environmental quality protection, sustainability of land ecosystems, and concerns of global climatic change, such efforts take on new significance and will be well received. We commend the organizers, authors, and editors for their efforts in providing geoscientists with these advances in modeling concepts.

LARRY P. WILDING, *president*
Soil Science Society of America

PREFACE

Studies of soil formation have traditionally relied heavily on qualitative or semiquantitative, functional models. These models have contributed greatly to our understanding of soil properties and their distribution in landscapes. Recent studies of ecological systems and their response to human impacts and environmental change utilize quantitative, mechanistic models to simulate the complex processes of plant growth and interactions between components of the soil–plant–atmosphere continuum. However, integrated mechanistic models of soil physical, chemical, and biological processes are scarce, and the soil is frequently treated as a black box. A perceived need to address this deficiency and to complement our dominant models of soil formation in light of the near completion of a once-over soil survey in the USA led to the organization of a symposium on "Quantitative Modeling of Soil Forming Processes." This symposium was organized by Division S-5, with Division S-9 as cosponsor, and was held in two sessions at the annual meetings of the Soil Science Society of America in Minneapolis, Minnesota, 7–12 Nov. 1992.

The objective of the first session of this symposium was to review the principles of quantitative, mechanistic modeling as applied to the environmental sciences and existing simulation models in the subdisciplines of soil physics, soil chemistry, and soil mineralogy which might serve as the building blocks for integrated physical, chemical and biological models of soil forming processes. In the second session, some pioneering efforts in the development of quantitative mechanistic models of soil forming processes were presented. Although we do not wish to detract from the importance of soil genesis studies using more traditional approaches we hope to foster a greater understanding of simulation modeling and perhaps inspire soil scientists and graduate students to further explore this approach to the study of soil forming processes. We also see quantitative, mechanistic modeling as a common tool and language across disciplines that will encourage collaboration within subdisciplines of soil science and facilitate linkage to other environmental disciplines in the future.

RAY B. BRYANT, *coeditor*
Department of Soil, Crop and Atmospheric Sciences
Cornell University
Ithaca, New York

RICHARD W. ARNOLD, *coeditor*
USDA-SCS, Soil Survey Division,
Washington, District of Columbia

CONTRIBUTORS

T.M. Addiscott   Soil Scientist/Modeler, IACR, Rothamsted Experimental Station, Harpenden, Herts ALS 2JQ, England

J. Bouma   Professor, Agricultural University, Wageningen, Department of Soil Science and Geology, Agricultural University, P.O. Box 37, 6700 AA, Wageningen, the Netherlands

Ray B. Bryant   Associate Professor, Department of Soil, Crop and Atmospheric Sciences, Cornell University, Ithaca, New York 14853

C. Vernon Cole   Senior Research Scientist, Natural Resource Ecology Laboratory, Colorado State University, Fort Collins, Colorado 80523

Sabine Goldberg   Soil Scientist, USDA-ARS, U.S. Salinity Laboratory, 4500 Glenwood Drive, Riverside, California 92501

Don W. Goss   Research Scientist, Blackland Research Center, Texas Agricultural Experiment Station, 808 E. Blackland Road, Temple, Texas 76502

Gerhard Grosskurth   Geochemist, Institute of Geochemistry, University of Göttingen, Goldschmidtrasse 1, 37077 Göttingen, Germany

Marcel R. Hoosbeek   Graduate Research Assistant, Cornell University, Ithaca, New York; present address and title, Research Associate, Department of Soil Science and Geology, Wageningen Agricultural University, P.O. Box 37, 6700 AA Wageningen, the Netherlands

J.L. Hutson   Senior Research Associate, Cornell University, Department of Soil, Crop, and Atmospheric Sciences, Ithaca, New York 14853

B.J. Irvin   Research Assistant, Institute for Environmental Studies, University of Wisconsin, Madison, B102 Steenbock Library, 550 Babcock Drive, Madison, Wisconsin 53706

R.G. Knox   Forest Ecologist, NASA/Goddard Space Flight Center, Code 9232 Biospheric Sciences Branch, Greenbelt, Maryland 20771

E.R. Levine   Soil Scientist, NASA/Goddard Space Flight Center, Code 9232 Biospheric Sciences Branch, Greenbelt, Maryland 20771

G.M. Marion   Research Physical Scientist, U.S. Army, Cold Regions Research and Engineering Laboratory, 72 Lyme Road, Hanover, New Hampshire 03755

Alex B. McBratney   Professor of Soil Science, Department of Agricultural Chemistry &
                              Soil Science, University of Sydney, Ross Street Building A03, NSW
                              2006, Australia

K. McSweeney   Professor of Soil Science, University of Wisconsin, Madison,
                              Department of Soil Science, 1525 Observatory Drive, Madison,
                              Wisconsin 53706

Dennis S. Ojima   Research Scientist, Natural Resource Ecology Laboratory, Colorado
                              State University, Fort Collins, Colorado 80523

William J. Parton   Senior Research Scientist, Natural Resource Ecology Laboratory,
                              Colorado State University, Fort Collins, Colorado 80523

David S. Schimel   Research Scientist, National Center for Atmospheric Research, P.O.
                              Box 3000, Boulder, Colorado 80307-3000

William H. Schlesinger   Professor, Duke University, The Phytotron, P.O. Box 90340, Durham,
                              North Carolina 27708

B.K. Slater   Research Assistant, University of Wisconsin, Madison, Department of
                              Soil Science, 1525 Observatory Drive, Madison, Wisconsin 53706

Donald L. Suarez   Research Leader, USDA-ARS, U.S. Salinity Laboratory, 4500
                              Glenwood Drive, Riverside, California 92501

Hans J.M. van Grinsven   Senior Soil Chemist, National Institute of Public Health and
                              Environmental Protection, Laboratory of Soil and Groundwater
                              Research, P.O. Box 1, 3720 BA Bilthoven, the Netherlands

S.J. Ventura   Assistant Professor, University of Wisconsin, Madison, Department of
                              Soil Science, 263 Soils, 1525 Observatory Drive, Madison, Wisconsin
                              53706

R.J. Wagenet   Professor, Cornell University, Department of Soil, Crop, and
                              Atmospheric Sciences, 235 Emerson Hall, Cornell University, Ithaca,
                              New York 14853

Lambert G. Wesselink   Graduate Research Assistant, Agricultural University, Department of
                              Soil Science and Geology, Agricultural University, P.O. Box 37, 6700
                              AA, Wageningen, the Netherlands; current address is at National
                              Institute of Public Health and Environmental Protection, P.O. Box 1,
                              3720 BA, Bilthoven, the Netherlands

L.P. Wilding   Professor of Pedology, Texas A&M University, Soil and Crop
                              Sciences Department, College Station, Texas 77843-2474

Conversion Factors for SI and non-SI Units

Conversion Factors for SI and non-SI Units

To convert Column 1 into Column 2, multiply by	Column 1 SI Unit	Column 2 non-SI unit	To convert Column 2 into Column 1, multiply by
Length			
0.621	kilometer, km (10^3 m)	mile, mi	1.609
1.094	meter, m	yard, yd	0.914
3.28	meter, m	foot, ft	0.304
1.0	micrometer, μm (10^{-6} m)	micron, μ	1.0
3.94×10^{-2}	millimeter, mm (10^{-3} m)	inch, in	25.4
10	nanometer, nm (10^{-9} m)	Angstrom, Å	0.1
Area			
2.47	hectare, ha	acre	0.405
247	square kilometer, km^2 (10^3 m)^2	acre	4.05×10^{-3}
0.386	square kilometer, km^2 (10^3 m)^2	square mile, mi^2	2.590
2.47×10^{-4}	square meter, m^2	acre	4.05×10^3
10.76	square meter, m^2	square foot, ft^2	9.29×10^{-2}
1.55×10^{-3}	square millimeter, mm^2 (10^{-3} m)^2	square inch, in^2	645
Volume			
9.73×10^{-3}	cubic meter, m^3	acre-inch	102.8
35.3	cubic meter, m^3	cubic foot, ft^3	2.83×10^{-2}
6.10×10^4	cubic meter, m^3	cubic inch, in^3	1.64×10^{-5}
2.84×10^{-2}	liter, L (10^{-3} m^3)	bushel, bu	35.24
1.057	liter, L (10^{-3} m^3)	quart (liquid), qt	0.946
3.53×10^{-2}	liter, L (10^{-3} m^3)	cubic foot, ft^3	28.3
0.265	liter, L (10^{-3} m^3)	gallon	3.78
33.78	liter, L (10^{-3} m^3)	ounce (fluid), oz	2.96×10^{-2}
2.11	liter, L (10^{-3} m^3)	pint (fluid), pt	0.473

Mass

Column 1 → 2	Column 1	Column 2	Column 2 → 1
454	gram, g (10^{-3} kg)	pound, lb	2.20×10^{-3}
28.4	gram, g (10^{-3} kg)	ounce (avdp), oz	3.52×10^{-3}
0.454	kilogram, kg	pound, lb	2.205
100	kilogram, kg	quintal (metric), q	0.01
907	kilogram, kg	ton (2000 lb), ton	1.10×10^{-3}
0.907	megagram, Mg (tonne)	ton (U.S.), ton	1.102
0.907	tonne, t	ton (U.S.), ton	1.102

Yield and Rate

Column 1 → 2	Column 1	Column 2	Column 2 → 1
1.12	kilogram per hectare, kg ha^{-1}	pound per acre, lb acre^{-1}	0.893
12.87	kilogram per cubic meter, kg m^{-3}	pound per bushel, bu^{-1}	7.77×10^{-2}
67.19	kilogram per hectare, kg ha^{-1}	bushel per acre, 60 lb	1.49×10^{-2}
62.71	kilogram per hectare, kg ha^{-1}	bushel per acre, 56 lb	1.59×10^{-2}
53.75	kilogram per hectare, kg ha^{-1}	bushel per acre, 48 lb	1.86×10^{-2}
9.35	liter per hectare, L ha^{-1}	gallon per acre	0.107
1.12×10^{-3}	tonnes per hectare, t ha^{-1}	pound per acre, lb acre^{-1}	893
1.12×10^{-3}	megagram per hectare, Mg ha^{-1}	pound per acre, lb acre^{-1}	893
2.24	megagram per hectare, Mg ha^{-1}	ton (2000 lb) per acre, ton acre^{-1}	0.446
0.447	meter per second, m s^{-1}	mile per hour	2.24

Specific Surface

Column 1 → 2	Column 1	Column 2	Column 2 → 1
10	square meter per kilogram, m^2 kg^{-1}	square centimeter per gram, cm^2 g^{-1}	0.1
1000	square meter per kilogram, m^2 kg^{-1}	square millimeter per gram, mm^2 g^{-1}	0.001

Pressure

Column 1 → 2	Column 1	Column 2	Column 2 → 1
9.90	megapascal, MPa (10^6 Pa)	atmosphere	0.101
10	megapascal, MPa (10^6 Pa)	bar	0.1
1.00	megagram per cubic meter, Mg m^{-3}	gram per cubic centimeter, g cm^{-3}	1.00
2.09×10^{-2}	pascal, Pa	pound per square foot, lb ft^{-2}	47.9
1.45×10^{-4}	pascal, Pa	pound per square inch, lb in^{-2}	6.90×10^3

(continued on next page)

Conversion Factors for SI and non-SI Units

To convert Column 1 into Column 2, multiply by	Column 1 SI Unit	Column 2 non-SI Unit	To convert Column 2 into Column 1, multiply by
Temperature			
$1.00 (K - 273)$	Kelvin, K	Celsius, °C	$1.00 (°C + 273)$
$(9/5 °C) + 32$	Celsius, °C	Fahrenheit, °F	$5/9 (°F - 32)$
Energy, Work Quantity of Heat			
9.52×10^{-4}	joule, J	British thermal unit, Btu	1.05×10^3
0.239	joule, J	calorie, cal	4.19
10^7	joule, J	erg	10^{-7}
0.735	joule, J	foot-pound	1.36
2.387×10^{-5}	joule per square meter, J m^{-2}	calorie per square centimeter (langley)	4.19×10^4
10^5	newton, N	dyne	10^{-5}
1.43×10^{-3}	watt per square meter, W m^{-2}	calorie per square centimeter minute (irradiance), cal cm^{-2} min^{-1}	698
Transpiration and Photosynthesis			
3.60×10^{-2}	milligram per square meter second, mg m^{-2} s^{-1}	gram per square decimeter hour, g dm^{-2} h^{-1}	27.8
5.56×10^{-3}	milligram (H_2O) per square meter second, mg m^{-2} s^{-1}	micromole (H_2O) per square centimeter second, μmol cm^{-2} s^{-1}	180
10^{-4}	milligram per square meter second, mg m^{-2} s^{-1}	milligram per square centimeter second, mg cm^{-2} s^{-1}	10^4
35.97	milligram per square meter second, mg m^{-2} s^{-1}	milligram per square decimeter hour, mg dm^{-2} h^{-1}	2.78×10^{-2}
Plane Angle			
57.3	radian, rad	degrees (angle), °	1.75×10^{-2}

Electrical Conductivity, Electricity, and Magnetism

To convert Column 1 into Column 2, multiply by	Column 1 SI Unit	Column 2 non-SI Unit	To convert Column 2 into Column 1, multiply by
10	siemen per meter, S m^{-1}	millimho per centimeter, mmho cm^{-1}	0.1
10^4	tesla, T	gauss, G	10^{-4}

Water Measurement

To convert Column 1 into Column 2, multiply by	Column 1 SI Unit	Column 2 non-SI Unit	To convert Column 2 into Column 1, multiply by
9.73 × 10^{-3}	cubic meter, m^3	acre-inches, acre-in	102.8
9.81 × 10^{-3}	cubic meter per hour, m^3 h^{-1}	cubic feet per second, ft^3 s^{-1}	101.9
4.40	cubic meter per hour, m^3 h^{-1}	U.S. gallons per minute, gal min^{-1}	0.227
8.11	hectare-meters, ha-m	acre-feet, acre-ft	0.123
97.28	hectare-meters, ha-m	acre-inches, acre-in	1.03 × 10^{-2}
8.1 × 10^{-2}	hectare-centimeters, ha-cm	acre-feet, acre-ft	12.33

Concentrations

To convert Column 1 into Column 2, multiply by	Column 1 SI Unit	Column 2 non-SI Unit	To convert Column 2 into Column 1, multiply by
1	centimole per kilogram, cmol kg^{-1} (ion exchange capacity)	milliequivalents per 100 grams, meq 100 g^{-1}	1
0.1	gram per kilogram, g kg^{-1}	percent, %	10
1	milligram per kilogram, mg kg^{-1}	parts per million, ppm	1

Radioactivity

To convert Column 1 into Column 2, multiply by	Column 1 SI Unit	Column 2 non-SI Unit	To convert Column 2 into Column 1, multiply by
2.7 × 10^{-11}	becquerel, Bq	curie, Ci	3.7 × 10^{10}
2.7 × 10^{-2}	becquerel per kilogram, Bq kg^{-1}	picocurie per gram, pCI g^{-1}	37
100	gray, Gy (absorbed dose)	rad, rd	0.01
100	sievert, Sv (equivalent dose)	rem (roentgen equivalent man)	0.01

Plant Nutrient Conversion

To convert Column 1 into Column 2, multiply by	Column 1 Elemental	Column 2 Oxide	To convert Column 2 into Column 1, multiply by
2.29	P	P$_2$O$_5$	0.437
1.20	K	K$_2$O	0.830
1.39	Ca	CaO	0.715
1.66	Mg	MgO	0.602

1 Simulation, Prediction, Foretelling or Prophecy? Some Thoughts on Pedogenetic Modeling

T.M. Addiscott

Rothamsted Experimental Station
Harpenden, Herts, England

Computer models describe processes as diverse as global warming and the generation of revenue from taxation. In soil science they have probably been used most to simulate flows of water and solutes and the transformations of C and N in organic material. Any process should in principle be amenable to modeling, but some are more readily modeled than others because they can be quantified more easily in numerical terms. Pedogenesis is not easy to model because it is a composite of several processes and because the degree of development of a soil profile is not something that is readily expressed by a set of numbers. Despite these problems, some modeling approaches have evolved and these were reviewed recently by Hoosbeek and Bryant (1992). The present paper has two purposes. The first part presents a few general concepts that are relevant to most forms of modeling and seeks to relate these to pedogenesis, while the second part takes up the idea (e.g., Runge, 1973) that thermodynamics provides a framework for pedogenetic modeling but suggests that an approach based on nonequilibrium thermodynamics (Prigogine, 1947; Katchalsky & Curran, 1967) may be more fruitful than one based on traditional equilibrium thermodynamics.

GENERAL CONCEPTS

Most of the concepts discussed here were included in a review paper published recently (Addiscott, 1993), so the emphasis is on relating them to modeling pedogenesis rather than on expounding them in full. Further details are in the review cited.

Models, Parameters and Data

Any kind of model is a *representation,* and a computer model is no exception. With any scientific model, however, there is the added dimension that the model also is a form of *hypothesis,* in the case of a computer model for the soil, an *extended hypothesis,* because of the complexity of the soil and the number of interacting processes with which we are concerned. Practically all models are simplifications. This is usually the reason why they are usable, but it is very important to keep in mind that a model is only a hypothesis and a simplified one at that. The ways in which various categories of models are useful are discussed later.

A *parameter,* according to one dictionary's definition, is a quantity constant in the case considered but varying between cases. Soil parameters usually vary from point to point within a field, but this problem can be met by saying that the assembly of values remains constant in the case considered. The use of the term *data,* on the other hand, carries no implication of constancy. The quantity of water held in a soil at a given hydraulic potential may well be a parameter for a model, but rainfall and evaporation are data.

All the foregoing remarks apply to pedogenetic models as much as to other types of model. Simplification, in particular, seems to be a key question, and success in this area of modeling seems likely to depend on the identification of appropriate simplified representations of pedogenesis as on any other factor.

Validation

In a discussion of how scientists work Popper (1959) set forward his *hypothetic-deductive* principle in which he argued that science advances through scientists forming hypotheses and testing them against observations that can refute them and then often reformulating and retesting the hypotheses. If a model is a type of hypothesis we need the means to test reliably whether the hypothesis is refuted, and as models become more and more part of public policy-making, it becomes more and more important that they are subjected to proper critical scrutiny. Whitmore (1991) has provided quantitative criteria for validating models together with some useful comments on some of the associated problems. These criteria are quantitative and statistical in nature and this probably limits their usefulness for testing pedogenetic models at least for the present. Whether any satisfactory criteria are possible for the qualitative testing of models seems open to question.

Classification of Models

Models vary appreciably in their nature and purpose, and Addiscott and Wagenet (1985a) attempted to provide a framework within which to classify them; this is shown in outline in Table 1–1. The main distinction is between *deterministic* models, which presume that any given set of events leads to a uniquely definable outcome, and stochastic models, which presume the outcome to be essentially uncertain and are structured accordingly. Within deterministic models there is a further distinction between *mechanistic* and *functional* models. Mechanistic implies that the model incorporates the most fundamental mechanisms of the process as understood at present, while functional, or less mechanistic, models usually contain some simplified concept of the process and make no claim to fundamentality. Stochastic models divide somewhat similarly between mechanistic models in which the stochasticity is introduced through the

Table 1–1.   Classification of models.

1.	Deterministic†
	Mechanistic
	Functional
2.	Stochastic
	Mechanistic with randomly selected distributed parameters
	Based on probability density functions
Other considerations: purpose, complexity, flexibility, transferability	
	Qualitative or quantitative‡
	Organizational hierarchy
	Level of information

†Addiscott and Wagenet (1985a).
‡Hoosebeek and Bryant (1992).

use of randomly selected distributed parameters, and models in which the process is described with the minimum number of parameters, each expressed as a probability density function. The purpose of the model is a further consideration. Some models are used mainly for *research* purposes, to explore hypotheses and expose areas of poor understanding while others function more as practical *management* aids. There is no unequivocal distinction, but it is generally the mechanistic models that are used in research and the simpler functional models in management. Other considerations include complexity, flexibility and transferability. With complexity we are concerned with the number of processes treated, the fundamentality with which each is treated and, ultimately, the amounts of computing power and computing time needed; while flexibility is a question of the rigidity of the boundary conditions, the breadth of applicability and the range of information needed by and supplied by the model. Transferability is best summed up by the question, "How easy is it for a person other than the developer to use the model— on a computer other than the one on which it was developed?"

This framework is only partially applicable to pedogenetic modeling, the distinction between deterministic and stochastic models in particular being too arcane for the present state of the art in this field. Hoosbeek and Bryant (1992) retained the distinction between mechanistic and functional models but added that between quantitative and qualitative models. They also added in their scheme the question of organizational hierarchy in terms of levels of information. Both additions are strongly relevant in the context of soil development.

Simulation, Prediction, Foretelling and Prophecy

One word that is used too freely about models is "predict," for which the dictionary definition is "foretell" or "prophecy." Not all modelers would describe their activities in such terms. "Predict" has its root in the Latin word *"prae,"* meaning "before," and "simulate" is probably a safer word for most modelers, because it lacks any implication that the modeling activity preceded the events modeled. These four words can be ranked in what they imply, as follows:

1. *Simulate:*   Obtain an acceptable fit to measured data.
2. *Predict:*   Provide simulation *before* outcome of measurements known.

3. *Foretell:*   Provide prior estimate of outcome *before* events occur. Or suggest likelihood of an event before it occurs. (The latter is a particularly effective demonstration of the power of a model.)

4. *Prophecy:*   Tell forth the likely consequence of a particular course of action. (based on the saying that the Old Testament prophets were not so much foretellers as forth tellers.) A good example of this can be seen in the effect of the "Nuclear Winter" modelers (Turco et al., 1983) in bringing home to politicians the realities of nuclear war.

Parameter Variability Issues

Virtually all measurements made on the soil show variability within a given area. If these measurements are used as parameters for models we need to be aware of the likely consequences of the variability. There are four main issues: (i) the extent of the variance, whether or not it shows spatial structure; (ii) the linearity of the model with respect to the parameters; (iii) the scale of the variability in relation to the scale of the modeling exercise and any data used for validation. These are all interrelated in a complex way and most of them will be important in pedogenetic modeling only where models incorporate relationships that are not linear in the mathematical sense. A linear function is one in which the second partial differentials of the function with respect to its variables are zero. If a computer model cannot be expressed as a single function, one way of assessing its linearity with respect to a distributed parameter is to run the model in two ways; first, taking account of the mean of the distribution only, and then taking account of the mean and the variance. The closer the two results, the more linear the model is with respect to the parameter. Addiscott and Wagenet (1985b) discussed ways in which models can be run taking account of parameter variance. These issues were reviewed recently elsewhere (Addiscott, 1993) and will not be discussed further here except to note that the issue of scale seems likely to be the most important in a pedogenetic context.

Models are sometimes used at a scale quite different from the one at which they were developed and validated, and some of the questions relating to this practice were discussed in the paper cited above. Pedogenetic models seem particularly susceptible to the problem of scale because they have to simulate a process, profile development, that is a composite of other processes that themselves occur at differing scales, and because one of the key factors in soil formation is landscape, which is a very large-scale entity. Hoosbeek and Bryant (1992) have begun to address this problem; it is, as they say, "an arduous task."

THERMODYNAMICS AND PEDOGENESIS

One way of introducing a mechanistic element into a model for pedogenesis is to consider the thermodynamics of some or all parts of the process. This necessitates allocating the soil to the correct thermodynamic system, there being three possibilities. An *adiabatic* system can exchange neither energy nor matter with its surroundings, a *closed* system can exchange energy but not matter, and an *open* system can exchange both. An adiabatic system is essentially a theoretical concept that can only be approximated in experiments, but a closed system is relatively easy to realize. Most systems in the biosphere, however, are open systems that exchange both matter and energy with their surroundings, and the soil is clearly an open system (Johnson & Watson-Stegner, 1987).

Runge (1973) described a model, based on concepts evolved by Ballagh and Runge (1970) and Smeck and Runge (1971), that explained profile development in terms of ordering brought about by the flow of water through the soil. His basic idea, derived from the First and Second Laws of thermodynamics was that entropy was lessened and the profile therefore made more ordered, by an input of energy from outside the system, and he considered water flow to be the principal source of the energy. The main problem with Runge's approach, identified by Hoosbeek and Bryant (1992), is that Runge used thermodynamics appropriate to closed systems, whereas the soil is an open system. Indeed Runge's use of water *flow* as the source of energy implies a flow of matter into the system, which therefore ceases to be a closed system, and it is not only water that flows into and out of the soil. Thus we need a form of thermodynamics that deals with flows of matter in open systems and this leads us to nonequilibrium thermodynamics rather than the traditional equilibrium thermodynamics on which Runge's approach was based. Runge was essentially describing a *closed system* that tended towards an *equilibrium* characterized by minimum energy and *maximum entropy*. Using nonequilibrium thermodynamics enables us to describe an *open system* that tends towards a *steady state* characterized by *minimum production of entropy* (Prigogine, 1947; Katchalsky & Curran, 1967). There is a clear conceptual difference between the two systems, particularly in that one tends to maximum entropy and the other to minimum production of entropy, and it should be helpful to explore whether the concepts of nonequilibrium thermodynamics can be applied usefully to pedogenesis. [The following section was written before the author became aware of the paper of Smeck et al. (1983) on this topic, which includes the mineralogical aspects of the topic, which are not discussed here.]

We need first to clarify the relationship between thermodynamic parameters and order. One key concept is that entropy is inter alia a measure of disorder or randomness, so for example, because the molecules of a solid have a much more ordered configuration than the corresponding liquid, melting a solid leads to an increase in entropy (Glasstone, 1947). An ordered state contains information, so that entropy is to some extent equivalent to information. Information also is related to work (in the thermodynamic sense), such that there is a two-way interchangeability, work being convertible to information in some circumstances and information to work in others. It follows, therefore that continuous work permits the self organization of systems, and this is the principle that underlies the ordering of the biosphere (Morowitz, 1970).

Thermodynamic work is done when energy in the form of heat is transferred from a source at a high temperature to a sink at a low temperature. Continuous work therefore requires effectively infinite isothermal reservoirs at high and low temperatures and these are provided by the sun and outer space respectively, as shown in Fig. 1–1. The whole process involves a flow of energy from the sun to

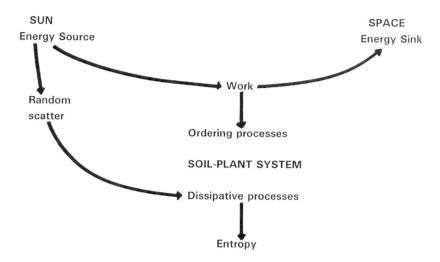

Fig. 1–1.   Performance of work and production of entropy in the soil–plant system.

outer space and therefore produces entropy, but the work done in local processes at the surface of the earth may lead to a great increase in order. Figure 1–1 shows that in addition to the ordering processes permitted by the continuous work there also are dissipative processes that result in the production of entropy at the local level as well as at the level of the solar system. These result from the random distribution of heat energy. The most obvious examples of these ordering and dissipative processes are biological. Photosynthesis and associated processes build complex, ordered biological structures containing substances of large molecular weight from small molecules such as CO_2, H_2O and NH_3, while dissipative processes tend to degrade these structures back into the small molecules. Morowitz (1970) has pointed out that it is the interaction between ordering and dissipative biological processes that drives the great ecological cycles of the biosphere. The soil has a key role in many of these cycles and they in turn influence the process of pedogenesis. The next section will therefore examine some of the main ordering and dissipative processes that affect the soil-plant system and soil genesis. Little is to be gained by considering the soil without the plant.

ORDERING AND DISSIPATIVE PROCESSES IN THE SOIL–PLANT SYSTEM

Table 1–2 shows a list of processes that seem likely to have either ordering or dissipative effects in the soil–plant system. They are listed in pairs, one ordering and one dissipative, although these are not necessarily exact opposites and they are grouped very broadly into biological and physical processes, the latter mainly involving water.

Table 1–2. Ordering and dissipative processes in the soil-plant systems categorized as biological or physical. The pairs are not necessarily exact opposites.

Ordering processes, entropy decreases	Dissipative processes, entropy increases
Biological	
Photosynthesis	Respiration
Growth	Senescence
Formation of humus	Decomposition of humus
Physical	
Water flow (profile development)	Water flow (erosion, leaching)
Flocculation	Dispersion
Aggregation	Disaggregation
Development of structure	Breakdown of structure
Larger units	Smaller units
Fewer of them	More of them
More ordered	Less ordered

Biological Processes

Photosynthesis is, as mentioned earlier, the principal biological ordering process, in which work is done converting CO_2 and H_2O into more highly ordered carbohydrate molecules. The corresponding dissipative process is *respiration* which converts part of the carbohydrate back to CO_2 and H_2O. Respiration is an example of a dissipative process that is facilitated by heat energy. Photosynthesis and respiration are of interest in the context of pedogenesis because their net outcome determines the amount of organic matter entering the soil.

Plant growth continues the ordering process initiated by photosynthesis and its associated assimilatory processes by converting the assimilated material into a plant comprising several organs and a variety of functions. *Senescence,* suggested as the corresponding dissipative process also is facilitated by heat energy, but it has an interestingly anomalous position in the present scheme because it is a dissipative process that is controlled to a substantial extent by *information* stored in the DNA of the plant. The most important aspect of growth in this context is root growth, which tends to be an ordering process with respect to the soil because it contributes to the development of the soil profile and because it enables the capture of nutrient ions which would otherwise be dissipated by leaching.

The formation of humus is listed as an ordering process, but there also is a dissipative aspect to it in that the soil, animals and microbes which convert organic material that enters the soil into humus convert some of this material into CO_2 and other small molecules. Overall, however, humus plays an important role in the ordering of the soil profile in two ways; humus formation is a part of the ordering of the profile itself, and humus acts as a cement that it contributes to soil aggregation and other aspects of profile stabilization. *Decomposition of humus* is, on balance, a dissipative process, producing CO_2, NH_4 and other small molecules, although one that contributes to the life processes of the soil biomass.

Physical Processes

Water is involved in several of the physical processes to be discussed. It conforms to the general scheme shown in Fig. 1–1 in that work is done when heat energy is used to overcome the latent heat of evaporation of water and that this energy is released, ultimately to space, when the water vapor condenses, usually at high altitudes. The resulting water in the liquid phase then has potential energy which becomes kinetic energy. Much of the latter is dissipated at the soil surface, but the water retains sufficient potential energy to flow through the soil.

Water flow can be an ordering process and contribute to *profile development,* as described by Runge (1973). It is, however, shown as both an ordering and a dissipative process in Table 1–1 because *erosion,* shown as the dissipative process corresponding to profile development, also is caused by water flow (and wind). Whether water flow proves to be an ordering or a dissipative process depends on whether or not the amount of kinetic energy applied per unit area by the rainfall at the soil surface exceeds some critical value (Morgan, 1986). This critical value depends on the soil type. Ordering of the soil profile also may be achieved when water is withdrawn from the soil by evaporation. Some of the Dutch Polders, areas of land reclaimed from the sea, have very heavy clay in places, and during reclamation the soil is gradually dried out so that cracks form in the clay through which water can drain and roots penetrate. Closely associated with water flow is solute *leaching,* which plays a part in profile development but which is a mainly dissipative process.

Flocculation, aggregation and the development of structure are all clearly ordering processes, but they involve subtle interparticle forces which are somewhat beyond the scope of this review and which have been discussed by others (Deryaguin & Landau, 1941; Emerson, 1983; Pashley & Israelachvili, 1984). The corresponding dissipative processes are self evident.

PEDOGENESIS AND ENTROPY PRODUCTION

Considering the concepts of nonequilibrium thermodynamics has enabled us to divide soil/plant processes into entropy-decreasing ordering processes and entropy-increasing dissipative processes. Can more be done? The main problem, analogous to that faced by Runge (1973), is that we cannot to any real extent *quantify* the entropic contributions of the various processes. One possibility might be to give arbitrary weights, based on informed guesses, to the contributions of the various processes on each side of the table, but the problems of assigning the weights would be exacerbated by the interactions between the processes that are discussed below. It may therefore be more productive to examine the theoretical treatment of nonequilibrium thermodynamics to see if it suggests any broad conclusions about entropy production in the soil-plant system.

Force-Flow Relationships and Entropy Production

The laws of Ohm and Darcy are familiar examples of relationships between flows and the forces on which they depend. Where a number of flows and forces operate in a system all the flows are influenced by all the forces. This principle, first enunciated by Lord Rayleigh for mechanical forces and flows in his treatise on sound, was extended to thermodynamic forces and flows [see Katchalsky & Curran (1969) for citation of Onsager (1931)] in a set of equations known as the phenome-nonological equations which relate flows, J_i, to their conjugate forces, X_i. If, for example, there are three forces and three flows the following equations result.

$$J_1 = L_{11}X_1 + L_{12}X_2 + L_{13}X_3$$
$$J_2 = L_{21}X_1 + L_{22}X_2 + L_{23}X_3$$
$$J_3 = L_{31}X_1 + L_{32}X_2 + L_{33}X_3$$

It is obviously important that the conjugate force is appropriate to the flow and this is ensured by the condition that the product of the flow and the force, J_iX_i has the dimensions of entropy production, or decrease in free energy, per unit time. This is analogous to the well-proven relationship that shows the heat produced when an electric current flows in a wire to be the product of the current and the electromotive force (EMF).

The coefficients L_{ij} that link the flows and forces are equivalent to the conductivity in the electrical analogue and are of two kinds, "straight" ($i = j$) and "coupling," or "cross-linking" ($i \neq j$). Onsager [see Katchalsky & Curran (1969) for citation of Onsager (1931)] was able to simplify the matrix by proving that where $i \neq j$, $L_{ij} = L_{ji}$.

These crosslinking coefficients provide a useful reminder that few flows occur in the soil without interacting with other flows. They also are important when we consider entropy production. Each flow with its conjugate force makes a contribution to entropy of J_iX_i per unit time, and these contributions are additive, so that the entropy production from the matrix above is

$$\sigma = J_1X_1 + J_2X_2 + J_3X_3$$

This relationship is not as simple as it may appear, because unless the coupling coefficients are zero, J_1 is determined by X_2 and X_3 as well as X_1, and similarly for J_2 and J_3. Thus interactions between flow processes contribute to the production of entropy.

The deduction that entropy production is influenced by interactions between processes as well as the processes themselves provides a warning against the oversimplistic use of thermodynamics in modeling pedogenesis and suggests that the theory outlined above cannot be used directly for that purpose. The theory, however, has two further consequences which may help to provide a framework for understanding the soil-plant system.

One of these consequences is the proof (e.g., Katchalsky & Curran, 1967) that, in an open system that is allowed to mature, entropy production will decrease with time and reach a minimum. The other, noted by the same authors, is that if a flow is perturbed by a change in its conjugate force, the flow will act to decrease the perturbation so that the system returns towards its original state. Their theory also implies that the perturbation will cause an increase in the production of entropy.

Some Inferences from the Theory

Katchalsky and Curran (1969) saw clearly the implications for processes in the biosphere of the theory outlined above. "There are several remarkable analogies," they wrote, "between an open system approaching a steady state and living organisms in their development towards maturity." They went on to suggest that the concept of minimum entropy production could be the physical principle underlying the evolution of the phenomena of life and to point out that living organisms have regulatory mechanisms that preserve the steady state by countering perturbations as in the system described above. What is true of individual organisms is arguably true of the ecosystems of which they are part. For a given set of forces, or constraints, an ecosystem will, over a period of time, mature to a particular steady state. The soil profile will initially be one of the constraints determining the direction in which the ecosystem matures but will remain part of it and will itself be modified during the maturing of the ecosystem. If the ecosystem is perturbed, its regulatory mechanisms will act to counter the perturbation and restore the steady state. The soil will be part of this process. One question that may arise is, when does a perturbation become so lengthy that the perturbed constraint has to be regarded as a new constraint leading to a new steady state. Another is the possibility of a catastrophic perturbation as a result of which the system is unable to redirect itself towards any steady state.

Let us seek some example of these processes. The clearest example of an ecosystem in a steady state is climax vegetation (probably better described as "steady state" vegetation) and the soil associated with it. In many parts of the world this is some form of forest, and the soil is an integral part of the ecosystem and would be totally different without the vegetation. Such systems can readily withstand temporary, localized perturbations. In traditional West African shifting cultivation, for example, the soil was cleared of the steady-state vegetation and cultivated for a few years, during which the fertility declined because of loss of organic matter. When yields became too small the area was allowed to revert to forest and the fertility was gradually restored (Nye & Greenland, 1960). Where the soil was virtually uncultivated and the crops were planted in holes made with "dibbling sticks," the soil profile was restored to its original form in about 10 yr, but if it was turned over with hoes the restoration took about 50 yr. The soil fauna played a vital role in the process (P.H. Nye, personal communication). Where the perturbation is neither temporary nor localized it may become a catastrophic perturbation as we have seen with the large-scale destruction of soil profiles in Amazonia.

Many areas of the world no longer have their steady-state vegetation but have not reacted catastrophically. The reason for this is gradual change. It may be no coincidence that the theory described above, which Katchalsky and Curran (1969) saw in effect as a model of biosphere processes, is not suitable for processes that are too rapid or too far removed from equilibrium. At Rothamsted Experimental Station we have very long-term data which permits comment on steady states and their perturbation on a gradual basis. The steady-state vegetation of the area that is Rothamsted is thought to be woodland. There have been two main forms of agricultural perturbation which probably happened in succession:

1. To very long-term grassland, which must have been cleared of most of the trees and kept free of them by grazing.
2. To arable land, cultivated for centuries.

We saw above that the perturbation of a steady state should in theory lead to an increase in entropy production. The transition from old grassland to plowed arable land certainly increases entropy production. Plowing up old grassland at Rothamsted led to the loss of 4 t ha^{-1} of N from the soil in the first 20 yr, and much of this could be found in the chalk beneath the site as nitrate. About 40 t ha^{-1} of C must have been lost as CO_2. The soil profile presumably became less ordered, but this probably made a smaller contribution to entropy production than the much enhanced production of small molecules. Considering the likely fate of the trees suggests that the perturbation from the steady-state climax vegetation to arable land, via grassland or directly, also must have brought an increase in entropy overall.

The theory discussed above not only shows that a perturbation of a steady-state system causes an increase in entropy production, but that when the perturbation is removed, the system will return towards the steady state with a decline in entropy production towards an eventual minimum. That this happens in ecosystems is suggested by two very interesting sites at Rothamsted. In the 1880s two sites which had long been in arable use were left uncultivated and remain so to the present day. The theory suggests that these sites should have reverted towards the steady-state vegetation. In 1957, Thurston (1958, p. 94) wrote of the more acid of the two sites, "The area has reverted to woodland, consisting chiefly of elm, ash and oak. The largest tree is an oak 81 inches (2060 mm) in diameter at 4 feet (1220 mm) from the ground, growing near the middle of the area that was cultivated. Of the 46 species of angiosperms present in 1957, thirty-two had been recorded previously and 14, including eight woodland species, had come in since 1913. In the same period, 55 species, all characteristic of grassland, had disappeared. All the arable weeds had already gone by 1913." The steady state had clearly been restored to a very large extent. Woodland also reasserted itself on the other site, but the calcarous nature of the soil there resulted in a differing distribution of species, with hawthorn (genus *Crataegus*) rather than oak (genus *Quercus*) predominant (Jenkinson, 1971).

With the restoration of the steady state should have come a decline in entropy production, and this clearly happened. By the time of Jenkinson's (1971) report, the acid site had accumulated 180 t ha^{-1} of trees and the calcareous site 274 t ha^{-1} of trees, both representing a massive ordering of small molecules, including, Jenkinson estimated, about 0.7 and 1.1 t ha^{-1} of N respectively.

More interesting in the present context is the corresponding ordering process in the soil. Between 1883 and 1964 the soils accumulated several tonnes of N and tens of tonnes of C per hectare (Table 1–3). The soil gained an average of 530 kg ha^{-1} of C and 45 kg ha^{-1} of N each year in the calcareous site and rather less in the acid site. This was not the only evidence of ordering—when the soil profile description in the calcareous site was compared with that in arable soil about

Table 1–3.   Increases of organic C and N, 1883 to 1964 in calcareous and acid soils allowed to revert to woodland (from Jenkinson, 1971; sampling depth-22.9 cm).

	Gain		Mean annual gain	
	Organic C	Organic N	Organic C	Organic N
	———— t ha⁻¹ ————		———— kg ha⁻¹ ————	
Calcerous soil	43.1	3.67	530	45
Acid soil	20.9	1.09	250	13

75 m away it showed evidence of greater development. This comparison needed to be made very cautiously, because the two descriptions were made on different occasions, but one person was involved in both descriptions and both used the terminology of the USDA Soil Survey Manual (Soil Survey Staff 1951).

The arable soil description is for Profile 1 of Avery and Bullock (1969, p. 63–81), while that for the site allowed to revert to woodland was made by Avery and King (1971, p. 134–135) and is published as the appendix to Jenkinson (1971). The two descriptions are shown side by side in Table 1–4. The main difference is that the arable soil was described as having weak fine and medium blocky structure between 0 and 20 cm and ill-defined coarse block structure below 20 cm, whereas the woodland soil was said to have moderately developed, mainly blocky structure, fine near the surface and changing to fine to medium, medium to fine and fully coarse with depth.

Table 1–4 loses information by taking the soil surface at each site as the reference datum. (The profile descriptions were not originally intended for comparison.) Choosing an independent datum would have shown more clearly that a key difference between the profiles lies in a stone-free layer of small bulk density to be found at the top of the woodland profile but not that of the arable soil; this results from earthworm activity. (D.S. Jenkinson kindly drew my attention to this point.) Thus there is a suggestion of a decline in entropy production due to profile development, but not one that can be quantified as well as the accumulation of C and N.

ENTROPY AND AGRICULTURE

Grazing and arable cultivation were discussed above as perturbations, albeit very long ones, of the steady state. The sites that were allowed to revert to the original steady state form only a small percentage at Rothamsted; most of the land continues in arable use or, to a much lesser extent, as grazed or cut grassland. These "perturbations" have now continued for centuries, and where they have been left unchanged they have effectively become constraints, so that the grass and arable land have themselves become steady states. Indeed, we have experiments at Rothamsted that have received the same fertilizer inputs for up to 150 yr, so that the inputs have become constraints and the plots getting them can be said to be in steady states. Steady-state plots with particular annual inputs of N from fertilizer were identified by Jenkinson (1991) as valuable resources for studying the flow of N in arable ecosystems and have yielded interesting results when used for ¹⁵N studies (Powlson et al., 1986).

Table 1–4. Comparison of soil profile descriptions for Broadbalk wilderness and Broadbalk field.

Broadbalk wilderness (wooded)†			Broadbalk field (cultivated) near wilderness‡		
Horizon	Depth	Description	Horizon	Depth	Description
	cm			cm	
A	0–10	Very dark grayish brown (10 YR 3/2) stony loam with moderately developed fine subangular blocky and granular structure, the latter best expressed in the first 2 cm; stones comprise small subangular flints and a few rounded flint pebbles; slightly plastic; nonsticky; friable; abundant fine fibrous and common small woody roots; earthworms throughout profile, slightly calcareous; merging boundary	Ap	0–20	Dark brown (10 YR 4/3–4/2) flinty clay loam to silty clay loam; friable to firm when moist, hard when dry and plastic when wet; irregular clods with weak fine and medium blocky structure; slightly calcareous (added chalk); few fine ferrimanganiferous nodules; few roots; sharp boundary.
Eb	10–30	Brown (10 YR 4/3) loam with moderately developed fine to medium subangular blocky structure and with inclusions of very dark greying brown (10 YR 3/2) more granular material from the surface horizon; stones as above, together with a few very small chalk fragments; nonsticky; slightly plastic; friable; common fine fibrous roots and occasional larger woody roots; very slightly calcareous; narrow irregular boundary.	IIBlt(g)	20–50	Yellowish red (5 YR 5/6–5/8) to strong brown (7.5 YR 5/6) flinty clay with common, fine to medium, red (2.5 YR 4/6) mottles and few, fine brown (7.5 YR 5/4) to pale brown (10 YR 5/3) mottles, chiefly in the lower part; smooth brown (7.5 YR 5/4) faces enclosing stones; very firm when moist, very hard when dry and very plastic when wet; ill defined (in core) coarse blocky structure; occasional channels partly filled with dark topsoil material; few diffuse ferrimanganiferous concentrations; roots rare; merging boundary.
Eb/Bt	30–40	Brown (7.5 YR 4/4) stony loam to clay loam; stones as above, up to medium in size; moderately developed medium to fine blocky structure; slightly sticky; slightly plastic; friable; few fine fibrous roots; very slightly calcareous; merging boundary.	II B2t(g)	50–95	Similar, yellowish red, strong brown and red, faintly mottled clay with fewer flints; common mottles and subvertical streaks of brown (7.5 YR & 10 YR 5/4–5/3) and (locally) light brownish grey (10 YR 6/2); very plastic when worked in the wet condition, but crumbles readily before moulding when partly dried; greyish inclusions are relatively plastic and apparently more clayey than the matrix; roots rare.
Bt(g)	40–60+	Reddish brown (5 YR 4/5) stony clay with common, faint, fine, paler brown and reddish mottling; stones as above; moderately developed coarse to medium blocky structure; sticky; plastic; firm; a few fine fibrous roots (note that the depth to this horizon varied from 30–75 cm in different holes).			

†B.W. Avery and P. Bullock, 1969, p. 63–81.
‡B.W. Avery and D.W. King, 1971.

Sustainable Agriculture

There has recently been much debate about how to make agriculture "sustainable" and there also has been a fair amount of debate about what "sustainable' means in this context. Natural ecosystems are sustainable not least because they become steady states characterized by minimum production of entropy. As we have seen, entropy production often involves the production of small molecules such as CO_2, N_2O, NO_3, and CH_4 all of which are regarded for one reason or another as environmentally undesirable when in excess. This suggests strongly that sustainability implies minimum entropy production and that it is agricultural systems that permit the establishment of steady states that we should be seeking. Continuous arable agriculture is often viewed with doubt, but it may have much to offer, provided that it does achieve a steady state and provided the flows in the system, particularly that of N, are of appropriate magnitudes. Organic farming, on the other hand, may need to be viewed more critically, because it usually involves the regular plowing up of grassland, so that neither the grassland nor the arable land achieves a steady state. It may be possible to approach a steady state in organic farming but this seems likely to be more difficult than in conventional agriculture. The balance of advantage may depend on the type and location of the soil. Organic farming may, for example, render some soils less susceptible to catastrophic perturbation, which may be particularly important in the Tropics.

CONCLUSION

Each day the soil becomes more important in supporting an increased world population and more threatened by its activities. Food production and environmental health are on a tightrope and there is no scope for error in managing the soil. The role of the pedogenetic modeler as prophet, "forth-telling" the consequences of political or management decisions cannot be overstressed.

The nonequilibrium thermodynamic model discussed above is, like all models, a simplification. The generation or regeneration of soil-plant systems depends on the biological potential as well as the thermodynamic capability. Soil fauna, for example, played a vital role in two of the examples of regeneration cited, and the concepts of the model proved applicable only because the biological potential was there. In the long term we cannot afford, in either our modeling or the management of ecosystems, to take the biological potential for granted.

ACKNOWLEDGMENTS

I am grateful to the Division S-5 group for the invitation to present this paper at their symposium on Quantitative Modeling of Soil Forming Processes, to the Soil Science Society of America for supplying my air fare and to D.S. Jenkinson and D.S. Powlson for their helpful suggestions. I am particularly grateful for Dr. Jenkinson's comments about biological potential.

REFERENCES

Addiscott, T.M. 1993. Simulation modelling and soil behaviour. Geoderma 60:15–40.

Addiscott, T.M., and R.J. Wagenet. 1985a. Concepts of solute leaching in soils: A review of modelling approaches. J. Soil Sci. 36:411–424.

Addiscott, T.M., and R.J. Wagenet. 1985b. A simple method for combining soil properties that show variability. Soil Sci. Soc. Am. J. 49:1365–1369.

Avery, B.W., and P. Bullock. 1969. Morphology and classification of Broadbalk soils. Rep. Rothamsted Exp. Stn., 1968. Part 2. Rothamsted Exp. Stn., Harpenden, England.

Avery, B.W., and D.W. King. 1971. Profile descriptions of wilderness soils. Rep. Rothamsted Exp. Stn., 1970. Part 2. Rothamsted Exp. Stn., Harpenden, England.

Ballagh, T.M., and E.C.A. Runge. 1970. Clay-rich horizons over limestone—illuvial or residual? Soil Sci. Soc. Am. Proc. 34:534–536.

Deryaguin, B., and L. Landau. 1941. Theory of the stability of strongly charged lyophobic soils and of the adhesion of strongly charged particles in solutions of electrolytes. Acta Physicochim. (URSS) 14:633–662.

Emerson, W.W. 1983. Interparticle bonding. p. 477–498. In Soils: An Australian viewpoint. Div. Soils, CSIRO, Melbourne, Australia/Acad. Press, London.

Glasstone, S. 1947. Textbook of physical chemistry. Macmillan, London.

Hoosbeek, M.R., and R.B. Bryant. 1992. Towards the quantitative modelling of pedogenesis—A review. Geoderma 55:183–210.

Jenkinson, D.S. 1971. The accumulation of organic matter in soil left uncultivated. Rep. Rothamsted Exp. Stn., 1970. Part 2. Rothamsted Exp. Stn., Harpenden, England.

Jenkinson, D.S. 1991. Rothamsted long-term experiments: Are they still of use? Agron. J. 83:1–10.

Johnson, D.L., and D. Watson-Stegner. 1987. Evolution model of pedogenesis. Soil Sci. 143:349–366.

Katchalsky, A., and P.F. Curran. 1967. Nonequilibrium thermodynamics in biophysics. Harvard Univ. Press, Cambridge, MA.

Morgan, R.P.C. 1986. Soil erosion and conservation. Longman Sci. Tech. Harlow, England.

Morowitz, H.J. 1970. Entropy for biologists. Acad. Press, New York.

Nye, P.H., and D.J. Greenland. 1960. The soil under shifting cultivation. Tech. Commun. 51. CAB, Harpenden, England.

Pashley, R.M., and J.N. Israelachvili. 1984. DLVO and hydration forces between mica surfaces in Mg^{2+}, Ca^{2+}, Sr^{2+} and Ba^{2+} chloride solutions. J. Colloid Interface Sci. 97:446–455.

Popper, K.R. 1959. The logic of scientific discovery. Hutchinson, London.

Powlson, D.S., G. Pruden, A.E. Johnston, and D.S. Jenkinson. 1986. The nitrogen cycle in the Broadbalk Wheat Experiment: Recoveries and losses of ^{15}N labelled fertilizer applied in spring and impact of nitrogen from the atmosphere. J. Agric. Sci. (Cambridge) 107:591–609.

Prigogine, I. 1947. Étude thermodynamique des processus irreversibles. Desoer, Liège.

Runge, E.C.A. 1973. Soil development sequences and energy models. Soil Sci. 115, 183–193.

Smeck, N.E., and E.C.A. Runge. 1971. Phosphorus availability and redistribution in relation to profile development in an Illinois landscape segment. Soil Sci. Soc. Am. Proc. 35:952–959.

Smeck, N.E., E.C.A. Runge, and E.E. Mackintosh. 1983. Dynamics and generic modelling of soil systems. p. 51–81. In L.P. Wilding et al. (eds.). Pedogenesis and soil taxonomy. I. Concepts and interactions. Elsevier, Amsterdam, the Netherlands.

Soil Survey Staff. 1951. Soil survey manual. USDA. U.S. Gov. Print. Office, Washington, DC.

Thurston, J.M. 1958. Geescroft wilderness. Rep. Rothamsted Exp. Stn. 1957. p. 94. Rothamsted Exp. Stn., Harpenden, England.

Turco, O.R., P.B. Toon, T.P. Ackerman, T.P. Pollack, and C. Sagan. 1983. Global atmospheric consequences of nuclear war. Science (Washington, DC) 222:1283–1292.

Whitmore, A.P. 1991. A method of assessing the goodness of computer simulation of soil processes. J. Soil Sci. 42:289–299.

2

Modeling Water and Chemical Fluxes as Driving Forces of Pedogenesis

R.J. Wagenet and J.L. Hutson

Cornell University
Ithaca, New York

J. Bouma

Agricultural University
Wageningen, the Netherlands

Interest in the soil forming process, much of which depends on water and chemical transfer, has existed since it was first recognized that soil processes play a key role in the survival of civilization. Since then, the recognition that the soil system was dynamic and regenerative, layered and subject to modification, and of different character spatially has aroused curiosity about the nature of soil forming processes. The interest persists today, and is manifested in continued soil investigations by scientists, engineers, and even social and political experts. Classical concerns of soil mineralogy, chemistry, classification and utilization have been augmented by expanded studies of how soil formation and degradation are being impacted by man's increasingly intense use and exploitation of the soil resource. This has resulted in models of pedogenesis based on physical and chemical processes. Simultaneously, conceptual models of soil formation have been developed as a framework within which characterization and taxonomic separation have been accomplished. Today, with environmental issues such as acid rain, climate change and waste disposal impacting the natural weathering and translocation processes that determine the nature of a soil profile, it is crucial that we are able to integrate and predict the magnitude and intensity of soil forming processes. These processes, the manner in which they vary at different temporal and spatial scales, and the formulation and application of models that describe water and chemical fluxes appropriate to each scale, need to be reexamined as they relate to pedogenesis. The first objective here is to briefly explain these modeling approaches and the types of relevant experimental methods that allow their use. The second objective is to place the models and methods into a hierarchial classification that separates them by temporal and spatial scale of application to issues of pedogenesis.

CLASSICAL CONCEPTS

One of the first documentations of interest in pedogenesis was the Russian emphasis in the late nineteenth century (Johnson et al., 1990) on soil classification for taxation purposes. This stimulus led directly to many subsequent investigations, first in Russia, then in Western Europe, and during the 1920s into North America, that established such concepts as the five soil forming factors (parent material, time, climate, organisms, relief) and soil zonality. The concepts of zonal soils that were derived from combinations of climate, parent material and subsequent weathering processes eventually was philosophically merged with the five factors model. This incorporation provided a framework within which soils could be grouped according to the conditions of their formation. These studies, qualitative beyond a few (now considered simple) measurements, gave impetus to the development of soil chemistry, mineralogy, physics and microbiology.

As important as these early studies were, they were produced in an environment that lacked both the philosophical and technological developments of the last half of the twentieth century. These early studies (e.g., Neustruev, 1927; Marbut, 1928; Jenny, 1941) were focused on soil characterization, using straightforward measurements of key static parameters. They were directed toward taxonomic separation and classification, generally with a recognition that soil is dynamic by nature, but characterizable only on a static basis, such as particle size separation, bulk density, color, and depths of observable horizonation. A number of taxonomies were developed, ever more fine and resolved, which were apparently perceived as inferential models of pedogenesis. These indirect approaches to modeling the evolution of soils have recently been complemented (not replaced) by the availability of new tools, such as hierarchy theory, simulation modeling, and advances in instrumentation for measuring dynamic processes in situ and the laboratory. It is important to review a few of these approaches as we consider future directions.

CONTEMPORARY FRAMEWORK

Processes of Soil Formation

Processes that lead to soil formation are a combination of weathering reactions and biological activity on a particular parent material, under given climatic and anthropogenic circumstances, with the products redistributed through the soil profile by water. Continuation of such processes over time and space results in the differentiation of soils in both the horizontal and vertical dimensions. Soil formation is a slow process, e.g., clay illuviation over 10,000 yr results in a textural B horizon, and podzolization occurs over at least a 500-yr span. The driving force of water in all its forms of rainfall, runoff, glacial advance/recession, streams, lakes, seas, groundwater and soil water is a fundamental component of the process of pedogenesis.

The effect of water is manifested across the landscape at different spatial and temporal scales, at different intensities at each scale, and with different consequences. Soils form and are destroyed by the effects of water at regional, landscape, pedon, profile and molecular levels. Water serves as a medium for chemical reactions and biological activity, and as a transporter of material in all three dimensions. Yet, our experience directly related to modeling water and chemical fluxes in nature is generally limited to much shorter times (as in pollution studies) and relatively smaller areas in space. Consequently, distinctly different models have been evolved, with many of them focused on rather narrowly defined components of the pedogenetic process. Modeling of the importance of water and chemical fluxes related to soil formation under such diverse circumstances is only possible where issues of spatial and temporal scale are first recognized.

The Consequences of Scale for Model Selection

Modeling of pedogenesis should be accomplished with recognition that the nature and complexity of the modeling process depends on the time period and areal extent being considered. Alternatively stated, the temporal and spatial scale of the modeling must be explicitly stated. The development and application of models to describe pedogenesis across multiple scales has recently been reviewed by Hoosbeek and Bryant (1993). They present an organizational hierarchy that allows models to be characterized based initially on ideas of Addiscott and Wagenet (1985) related to model categorization, but expanded to consider a hierarchy which describes the level at which a modeling approach aims to simulate the process of pedogenesis.

All models, including pedogenetic models, can be characterized in several ways. In one case, models are defined by the community of model users, such as the recognition of research, teaching, management or regulatory models. In another, the use of analytic or numerical solutions is the appropriate criteria. Perhaps more useful in the present case is a threefold characterization with respect to relative degree of computation, complexity, and level of organization (Hoosbeek & Bryant, 1993). The first characteristic, "degree of computation," distinguishes between qualitative and quantitative models and includes not only computer codes, but also mental models, consisting of concepts and ideas that exist only in the human mind, and which enable a soil scientist to work within the complex natural soil system (Dijkerman, 1974). From this conceptual framework arise verbal models expressed in spoken or written language. Mental, verbal and descriptive models are qualitative and are placed at one extreme of the continuum. Mathematical models attempt to formalize these abstractions as algorithms using either a deterministic or a stochastic approach, both of which are defined here as quantitative. Deterministic models are based on the premise that a particular set of characterization and input information will produce one uniquely defined model prediction. Multiple executions of deterministic models can be used in an attempt to represent heterogeneities and variabilities in soil processes (e.g., Petach et al., 1991; van Lanen et al., 1992). By contrast, a stochastic model presupposes that input information and the processes modeled can only be estimated within certain statistical limits, and therefore the model prediction must have a statistical uncertainty as well (Addiscott & Wagenet, 1985). Quantitative models as contrasted to mental, verbal or written models are placed at opposite extremes of the "degree of computation." Models used to estimate the effect of water as a driving force in pedogenesis fall somewhere between these two extremes.

The second characteristic, complexity of model structure, distinguishes functional and mechanistic models. Functional models use simplified or empirical representations of basic process to accomplish their predictions. This simplifies the model structure, reduces needed input data, and reduces computational time. These models do not, however, allow one to learn much about system behavior in terms of basic process. These models generally depend either on statistical relationships such as regression equations, or on simplified, empirical conceptualizations of water and chemical movement characterized by static "capacity" terms, such as field capacity, saturated water content, and bulk density (e.g., Marion et al., 1985). Alternatively, mechanistic models incorporate fundamental mechanisms of the processes involved (at least best as they are known!). Their degree of complexity corresponds to the model developer's concept of the present state of scientific knowledge (Addiscott & Wagenet, 1985). These are generally deterministic models that recognize that the dynamics of the system depend on rate processes, such as the rate of water flow, or the kinetics of mineral dissolution. These models are more complex because they require the use of differential equations and iterative procedures to provide solutions. Such models are used to test hypotheses or as tools to explore less well-understood areas of knowledge, which makes them inherently more complex.

The third distinction is based on the organizational hierarchy, which describes at which level a model aims to simulate a natural system (Table 2–1). As described by Hoosbeek and Bryant (1993), each level can be regarded as a system by itself, with its own terminology, and can be seen as a combination of subsystems at lower levels or as a subsystem of higher level systems. Each level integrates the knowledge of subsystems at lower levels, which means that investigations at a subsystem level, e.g., i-1, provide a mechanistic understanding of a model at the i-level. In the case of pedogenesis, the pedon is placed at the central i-level in the hierarchy. Soil Taxonomy (Soil Survey Staff, 1992) defines a pedon as a three-dimensional natural body large enough to represent the nature and arrangements of its horizons and variability. The other levels (modified after Dijkerman, 1974) were defined similarly, i.e., three-dimensional bodies large enough to represent the nature and extent of processes relevant at that level (scale), and the variability in them. Investigations of process and variability must then be investigated at a level of resolution appropriate for the level at which a model aims to simulate the system. The level of resolution being used depends on the scale (i-level) at which a model aims to simulate, the soil properties used in the model, and the measurement methods. This will be expanded in more detail below.

The three distinguishing characteristics just described may be combined and depicted graphically (Fig. 2–1) as axes in a three-dimensional framework for classification (Hoosbeek & Bryant, 1993). Both "degree of computation" and "complexity" are continuous functions with models positioned somewhere along the axes. The hierarchial levels are divided into "positive i-levels" (i, $i + 1$, $i + 2$ ...$i + 6$) and "negative i-levels" ($i - 1$, $i - 2$...$i - 4$) along the vertical axis. Groupings of models for further discussion are rather artificial, especially for

Table 2–1. Scale, modeling approaches and measurement methods relevant to description of the effects of water and chemical fluxes on pedogenesis.

Scale	System	Examples of appropriate modeling	Appropriate measurement or estimation methods†
$i+6$	World	Conceptual	Remote sensing, climate
$i+5$	Continental	Conceptual	Remote sensing, climate
$i+4$	Counties, states provinces	Statistical models	Multiple field sampling of soil characteristics, aerial photography
$i+3$	Soil region (interacting watersheds)	Hydrological models Mass balance models (stochastic, statistical, deterministic/functional)	Geohydrological techniques, fuzzy clustering
$i+2$	Catena or watershed	Catchment models, distributed or statistical hydrologic models (mixture of deterministic and stochastic)	Geostatistics, geohydrological techniques (hydrograph, stream chemistry)
$i+1$	Polypedon (field)	Two- or three-dimensional, lateral flows [ad-hoc stochastic use of $(i+1)$ models] (deterministic/functional)	TDR, GPR, geostatistics
i	Pedon	Mass flow modeling $(i-1)$ models applied with knowledge of variability (deterministic/functional, descriptive)	TDR, neutron attenuation
$i-1$	Profile horizon	One- or two-dimensional deterministic leaching models; pattern recognition (deterministic/mechanistic, descriptive)	Tensiometers, resistance blocks
$i-2$	Secondary structures (peds, aggregates)	Bypass flow, preferential flow macropores (deterministic/mechanistic, descriptive)	Soil peels, dye, fiber optics, CT
$i-3$	Basic structures (grain interactions)	Flow in basic fabrics (deterministic/mechanistic)	Thin sections, stereology, NMR
$i-4$	Molecular (pore/particle)	Electrochemical modeling (deterministic/mechanistic)	Submicroscopic techniques, solution chemistry

†GPR = ground penetrating radar, TDR = time domain reflectometry, CT = computer-assisted tomography, and NMR = nuclear magnetic resonance.

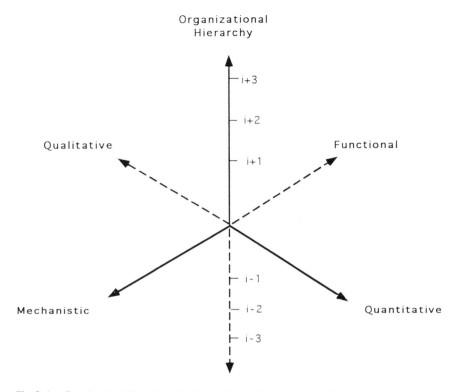

Fig. 2–1.   Organizational hierarchy of pedogenetic modeling approaches (Hoosbeek & Bryant, 1993).

models with multiple characteristics (e.g., a model that is partly descriptive and partly mathematical). Yet such a discussion illustrates clearly the modeling and measurement approaches that should be adopted at each scale as the effects of water and chemical fluxes on pedogenesis are considered.

It should be recognized from the outset, that unless soil physical and chemical processes are well described, modeling of soil water and chemical flux by whatever methods are going to be inaccurate, particularly at negative i-levels. The scale at which these processes are studied or summarized must correspond with the scale of the modeling effort. At higher i-levels, models that simulate these processes need to be very sensitive to the type of experimental method being used to develop characterizing relationships. Hydrodynamic factors of flow through porous media and the condition of the mineral surface (freshness, roughness, presence of coatings) are major factors in the discrepancies between laboratory experiments of different researchers and between laboratory and field conditions. Yet, perhaps in the final analysis both types of observations are real and accurate, and the only difference between them is the result of the scale at which they have been made. This is an issue that should be addressed in future studies.

Modeling Water and Chemicals in a Pedogenetic Context

Consideration on an individual basis of the several scales at which pedogenesis occurs (Table 2–1) is necessary to identify the appropriate issues to be understood at each scale, and the measurement or estimation methods best used (Table 2–2).

At the $i - 4$ scale, molecular level water and chemical interactions are studied in a deterministic, mechanistic manner. Understanding of the basic nature of particle–particle interactions, swelling and shrinking phenomena and the influence of pore and particle size on the water and chemical flow regime are considered. Soil chemists, mineralogists, and physicists have traditionally been involved in such studies, which now utilize sophisticated instrumentation such as submicroscopic techniques and nuclear magnetic resonance (NMR). In these cases, modeling of water and chemical fluxes is between the pore solution and the particle surface with very local variations in chemical potential energy being the main driving force for redistribution of water and chemical. Fluxes at this level are the essence of the soil forming process, with more macroscopic (higher i-level) forces providing the mediating driving variables that produce different soils temporally and spatially.

The $i - 3$ level is governed by flow in pores between the elementary particles which form the basic fabric (Brewer, 1964). Fluxes correspond to matric potential energies between zero and approximately -100 cm. Forces related to surface tension and contact angle are combined with the adhesive and cohesive forces described at the $i - 4$ level. This three-dimensional network allows consideration of the basic elements of unsaturated flow in the soil matrix as a process that leads to chemical redistribution. At this level, only one three-dimensional unit (a ped) is being considered, or a fragment of an apedal soil matrix with a more or less constant basic fabric and without macropores. Bouma and Anderson (1973) reported hydraulic conductivity (K) and moisture retention data of six basic fabrics ranging from sand to clay, demonstrating a decrease of K_{sat} in the increasing clay content and a corresponding increase of K_{unsat}. Bouma and Denning (1974) showed that soil morphological methods could be used to calculate hydraulic characteristics from morphometrically determined pore sizes in sands, which was not possible in more clayey materials with finer pores. Recently, the NMR technique has been used to characterize fluxes in small soil volumes in a nondestructive manner (Hinedi, et al., 1992), and FTIR (Fourier Transform Inra-Red) used to study frozen clay-water systems (Mizoguchi, 1992).

Secondary structures consisting of peds or of apedal soil material in the macropores define the $i - 2$ level. Two-domain systems of water and chemical fluxes are distinguished here according to flow through the soil matrix or between peds. These voids are variously termed macropores, preferred pathways or channels to indicate regions where the driving force for water and chemical fluxes is gravitational potential energy. Flow through the matrix is considered a separate, but hydraulically linked, domain. Experiments in the laboratory and field have demonstrated that both antecedent water content and subsequent boundary conditions also play a large role in the proportion of matrix and macropore flow (e.g., Hoogmoed & Bouma, 1980).

Table 2–2. Example references of modeling of water and chemical fluxes focusing on pedogenesis, or using models that could be focused on pedogenesis.†

Scale	Citation	General description of modeling†
$i + 6$	Smeck et al. (1983)	Graphical representations of entropy changes leading to development of soil orders
$i + 5$	Jenny (1980)	Framework presented for soil formation across multiple scales
$i + 4$	Shovic & Montagne (1985)	Statistical model of soil-landscape relations
	Havens (1988)	GIS-based statistical model of percentage of soil surface covered by rock fragments
	Lee et al. (1988)	Thematic map data, topography, digital elevation used to determine soil characteristics
$i + 3$	Petach et al. (1991)	Mass balance leaching model, GIS combined to estimate water and chemical fluxes
	Delcourt & Delcourt (1988)	Soil formation as a component of landscape analysis and ecosystem
$i + 2$	Marion et al. (1985)	Compartment model that integrates chemistry, ET, soil-water fluxes
	Wosten et al. (1990)	Extrapolation of point methods to areas using morphological/hydraulic functions
	Evans & Roth (1992)	Model concepts used to develop information for simulation modeling
	Bouma & van Lanen (1987)	Pedotransfer functions to estimate dynamic processes from static characterization data
$i + 1$	Finke (1993)	Mechanistic modeling of nutrient fluxes based on interpolated point data using disjunctive cokriging
i	Levine & Ciolkoz (1986)	Two-horizon model used to screen soils for sensitivity to acid deposition
	Hutson & Wagenet (1992)	Deterministic, mechanistic, comprehensive modeling of water and chemical fluxes
	Brimhall et al. (1991)	Cyclical dilational mixing used to estimate profile differentiation
	Kirkby (1985)	Weathering, organic and inorganic profiles simulated by selected dynamic processes
	Chadwick et al. (1990)	Mass balance functions used to estimate beach sand conversion into Alfisol
$i - 1$	Anderson & Bouma (1973)	Hydraulic conductivity of argillic horizon from analysis of interpedal pore patterns
	Bootlink & Bouma (1991)	Estimation of flow pathways using dye patterns
$i - 2$	Vepraskas & Bouma (1976)	Fe movement in homogeneous soil matrix with macropores, due to reduction/oxidation
	Peyton et al. (1992)	X-ray CT to measure macropore diameters
$i - 3$	Bouma & Anderson (1973)	Pore patterns in thin sections, stereology
	Bouma & Denning (1974)	Morphometric analysis of hydraulic conductivity
$i - 4$	Manley et al. (1987)	Physicochemical models describing fluxes between pore solution and particle surfaces
	Stumm et al. (1985)	Effects of complex ligands on mineral dissolution

†GIS = geographic information systems, ET = evapotranspiration, and CT = computer assisted tomography.

Modeling at this scale is the focus of much current activity but has produced mixed results. Descriptive models have been developed based on dye studies and limited in situ measurements, but as yet have not provided the linkage between morphology and soil hydrology necessary for information developed at this scale to be directly used at the $i - 1$ scale. It also has led to insufficient quantification of the effects of macroporosity on the fluxes of water and chemicals during pedogenesis studies focused at this scale. Soil morphological features are often helpful to characterize flow patterns in two-domain systems. Iron coatings on the walls of macropores may indicate dominant flow of water through the soil matrix and movement of reduced Fe to the faces of the peds where it is oxidized. Bleached outer zones of peds may indicate dominant flow along the macropores, causing reduction along the edges of the peds and oxidation within the soil matrix as the water infiltrates into the peds (Vepraskas & Bouma, 1976). While the effects are obvious, mechanisms and models are only partially understood.

The $i - 1$ level, the profile horizon, has been the most studied component of the pedogenetic hierarchy. Most taxonomy has been based on the extrapolation of $i - 1$ diagnostic horizons to the i-level or greater. Functional interpretation of major horizons in terms of physical, chemical and biological properties has been used as a basis for interpolation techniques while "depth to" and "thickness of" are used as attractive continuous regional variables (e.g., Bouma, 1989). Hydraulic conductivities of an argillic horizon were calculated based on flow patterns between peds as observed with morphological techniques (Anderson & Bouma, 1973).

Measurement techniques at the $i - 1$ level have substantially improved in the last two decades. An important development was the adoption of the representative elementary volume (REV) as a sampling volume necessary to define for each soil the number of appropriate peds needed for an accurate in situ estimation of dynamic processes (e.g., Lauren et al., 1988). Once the REV is determined, estimation of the influence of water and chemical fluxes on pedogenesis at the horizon scale should then be both modeled and measured at the size of the REV. Accurate estimation of soil hydraulic properties (hydraulic conductivity, water content, matric potential relationships), and their translation to other, insufficiently characterized locations through pedotransfer functions have focused on issues at the $i - 1$ scale (e.g., Bouma, 1989). Future studies related to issues of water and chemical fluxes and their effects on pedogenesis at the soil horizon level, must consider identification of appropriate REVs before modeling or measurement is undertaken.

The pedon (i-level of the hierarchy) further defines the lateral dimension of the pedogenetic description. As a three-dimensional lateral and vertical body, the pedon is defined to be large enough to represent the nature and arrangements of its horizons and variability. By use of the word "variability" the system is now recognized to be composed of a dominant soil pattern of horizonation, but with inclusions or variations in this horizonation. Spatial and temporal variability of processes within a pedon now need to be described for an accurate assessment of water and chemical fluxes as pedogenesis is considered at this scale. Spatially and temporally variable boundary conditions and consideration of vegetation, crops, climate and groundwater are now needed (Wagenet et al., 1991).

   The relative abundance of water and chemical transport models specifically developed or available to describe soil development confirms the interest at the pedon scale. Modeling approaches vary from deterministic/mechanistic water flow and chemical leaching models to new (statistical/conceptual) approaches based on pattern recognition, as in the case of dye traces and mottling patterns (Anderson & Bouma, 1973; Bootlink & Bouma 1991). The leaching models at this scale utilize both the principles of water and chemical transport and the principles of solution chemistry to combine physical and chemical systems. The determinism of leaching models at this scale allows the incorporation of temporally variable boundary conditions, such as rainfall, temperature and wetting/drying cycles in a more explicit representation of likely field conditions (Hutson & Wagenet, 1992). This provides the opportunity to test not only the sensitivity of the models, but also our hypotheses about basic process and process integration.

   Dynamic simulation models are not new, but vary tremendously in terms of being comprehensive and mechanistic. Most are quantitative but somewhat limited in scope. Yet they can perhaps be used as tools to translate results from one scale to the next. This is true in many cases of modeling, not just for those models used for description of water flow and chemical transport during pedogenesis. For example, Kirkby (1985) developed a mathematical model for soil profile development ($i - 1$ level) that considered a weathering profile, organic profile, and inorganic profile simulated by considering percolation, equilibrium solution, leaching, ionic diffusion, organic mixing, leaf fall, organic decomposition, and mechanical denudation. The approach describes the "proportion of substance remaining," p, at any depth. The accumulated deficit of weathered material is

$$w(z) =_{z=0} \int^{\infty} (1-p)dz$$

where z is depth below the soil surface. The total flow below depth z in the soil, the maximum percolation at depth and the process of weathering are estimated from knowledge of hydraulic properties. This approach neglected processes such as physical translocation of clay, complexing, chemical translocation, ion exchange and adsorption phenomena. The solution chemistry is based on the assumption of equilibrium with the mineral phase (not always true). Given such assumptions, and the simplification of basic process inherent in this type of model, not much is learned about basic processes in this approach. However, this model demonstrates a valuable connection between the soil profile and adjacent landscape and, despite the simplifications, provides a basis for model translation from the horizon ($i - 1$ level) to the pedon (i-level) or larger areas.

   Other models of water and transport in the context of describing pedogenesis at the i-level exist, focused on different situations. Compartment models, which are functional and semiquantitative, include the work of Levine and Ciolkosz (1986), Mayer (1985), and Chadwick et al. (1990). The Chadwick model presented a methodology, referred to as "a mass balance interpretation of pedogenesis," to estimate rates of weathering and pedogenesis over long periods of time. A set of analytical mass-balance functions consisting of basic conservation equations, strain equations (to account for volumetric changes), and flux equations were combined with traditional selective extraction and particle-size separation procedures to investigate the overall long-term pedogenetic processes that turned beach sand into an Alfisol.

These models provide quantitative estimates of water and chemical fluxes that could be very useful in pedogenetic studies. As yet, their accuracy for these purposes is not well established. This may be partially due to a lack of understanding of basic process, which should come from future studies at lower i-levels. However, the nature of the measurements used as input to the models or to test models also is important. Experimental techniques useful at this scale include those in which multiple measurements can be repeatedly and economically made. These methods include time-domain reflectrometry (TDR), neutron attenuation, and automated tensiometric systems.

The $i + 1$ level, the polypedon, is a field scale situation. As a collection of pedons, the polypedon represents a further lateral extension of the pedogenetic process. At this scale the use of two- or three-dimensional models that include lateral flow provides the ability to describe not only leaching, but also runoff and erosion, as well as subsurface processes of interflow and groundwater movement. Spatial variation of processes and properties at this scale is tremendously increased, and the usefulness of deterministic modeling of water and chemical fluxes as estimators of pedogenesis begins to diminish. Ad-hoc stochastic use of deterministic models through Monte Carlo-type methods that recognize the existence of variability, and attempts to treat it with a combination of mechanistic or functional relationships and statistical analysis (e.g., Wagenet et al., 1991; Finke & Stein, 1993) are necessary to investigate pedogenesis at this scale.

The evolution of positive i-level mechanistic models of water and chemical transport has been reviewed elsewhere (e.g., Wagenet, 1990). Beginning with the work in the early 1970s of Nimah and Hanks (1973), several other models have been developed that are potentially useful in pedogenetic studies at scales from the $i - 1$ to $i + 1$. The basic premise is that the models must describe both water and chemical transport, and must also be of sufficient flexibility to accommodate inclusion of relevant chemical equilibrium calculations. Although none of these has yet been applied at any i-level to treat pedogenesis, they represent a great potential for advancement of understanding of the process. For example, the LEACHM (Leaching Estimation And CHemistry Model; Hutson & Wagenet, 1992) contains such elements. It uses a numerical solution to the Richards equation dependent on knowledge of soil hydrological characteristics (K-θ-h relationships), boundary conditions, and source and sink terms to predict water flow. Subroutines are provided to estimate water retentivity and hydraulic conductivity parameters from particle-size distribution data; water fluxes are calculated considering daily potential evaporation and transpiration, and infiltration from irrigation and rainfall. Once the water fluxes are estimated, the corresponding chemical fluxes can be estimated using a numerical solution to the diffusion-convection equation, taking into account concurrent sources and sinks of solute (such as chemical equilibrium calculations that consider precipitation/dissolution reactions) and multiple ion exchange or sorption on the solid phase. The flexibility of the structure of the subroutines would allow new processes such as Al chemistry, dissolution kinetics, metal ion solubility, pH, CO_2, acid rain or horizonation to be included. The general theory of this approach, and its application to issues of solute migration (inorganic salts, fertilizers, pesticides) has been presented in several publications (e.g, Wagenet, 1983; Wagenet & Rao, 1990; Hutson & Wagenet, 1992).

Multiple measurements are required for characterization, and spatial statistical techniques such as geostatistics or temporal statistics such as time series analysis become more important at the $i + 1$ level. Relevant results at the field scale can be obtained when simulations are made for well-characterized point locations and when data are interpolated with techniques that allow expressions in terms of probabilities of occurrence (e.g, Finke & Stein, 1993).

The catena or watershed level $(i + 2)$ scale demands that the effects of water and chemical fluxes on pedogenesis be translated upward to a landscape level (Richardson et al., 1992). At this scale mechanistic and deterministic models can still be used when spatial patterns are described in terms of depth and lateral extension of profile horizons $(i - 1$ level). Geostatistical techniques can be used for interpolation of point calculations but preferably within homogeneous subareas (e.g., Stein et al., 1991), while basic physical and chemical data are obtained for each horizon by measurement or estimation (e.g., Wosten et al., 1990).

As we move our scale of concern to the polypedon or catena scale, it should be recognized that there are a number of reasons for quantitative mechanistic, positive i-level models of soil-water and chemical movement beyond the desire to describe and interpret pedogenesis. For example, ecological studies occur at the positive i-levels. The soil compartment in such studies is often the least understood and least quantified part of the ecosystem and the soil is frequently (mis)represented as a black-box between the more quantitative mechanistic models of climatologists, biologists, and hydrologists who lack appreciation of soils. Quantitative, mechanistic, pedogenetic models are needed to interface with ecological models, and as well for the education of other professionals as to the importance of soil. Such soil/water/chemical transport models at positive i-levels serve to extend basic soil processes into descriptions useful in understanding how soils function in a changing environment.

A particular example illustrates this point. Marion et al. (1985) developed a regional soil genesis model for $CaCO_3$ deposition in desert soils (CALDEP) that consists of five major components: a stochastic precipitation model based on monthly data resulting in three precipitation seasons; an evapotranspiration model; chemical thermodynamic relationships of the carbonate system; soil parameterization (pCO_2 per horizon, water-holding capacity); and soil water and $CaCO_3$ fluxes (only saturated flow was considered, influx of Ca is through weathering and from dust and precipitation). The model was executed using the present climate and three Pleistocene climate scenarios and was highly sensitive to the frequency of storm events, water holding capacity, and biotic control of pCO_2. Although such models have been developed and used before, the CALDEP model is one of the first soil models that used several component models to specifically simulate a soil forming process. Each component model borrowed knowledge from other disciplines (statistics, climatology, soil physics, thermodynamics, soil chemistry, and soil characterization).

At the $(i + 3)$ level of interacting watersheds, models of water and chemical fluxes become less mechanistic and deterministic, and more functional and statistical (e.g., Bouma et al., 1980). Physically based models that use process-level chemical and physical relationships are not as useful at this scale of variation, and the process representation becomes embedded in lumped parameters that are used at the scale of square kilometers. The lumped parameter models provide estimates of runoff, erosion and stream flow, but only infer the basic physical and chemical processes related to these predictions. It was at these scales that most initial studies in the nineteenth and early twentieth century were established, and it is at these scales that landscape ecologists still operate today.

An example of the position of pedogenesis (at our $i + 2...i + 4$ scales) in the hierarchy of landscape ecologists is revealing if only for the perspective it supplies. Two figures (Fig. 2–2 and 2–3) from Delcourt and Delcourt (1988) provide examples. In the first figure (Fig. 2–2), a hierarchial characterization of spatial and temporal domains intended for use by landscape ecologists represents soil formation as a microscale process, with the operative geomorphic processes at this scale of their hierarchy including soil creep, movement of sand dunes, debris avalanches, slumps, fluvial transport and deposition. Changes in the landscape mosaic are emphasized, and the modeling of pedogenesis, in the eyes of a soil scientist, is reduced to the treatment of rather macroscale events. This perception is somewhat tempered in Fig. 2–3, in which soil formation is recognized to occur at spatial scales of 10^0 to 10^{12} m^2 and 10^0 to 10^4 yr (or greater). Placed in context by landscape ecologists, soil formation is in fact more closely represented as it would have been by soil scientists 80 yr to 100 yr ago. In such cases, soil formation is a consequence of numerous natural processes, which are expressed at multiple interacting temporal and spatial scales. The effect of water and chemical fluxes on soil development at such large scales can only be expressed in broad terms, as is the effect of any process. In these cases the complexity of processes impacting soil formation argues for a broadly defined model of pedogenesis, more holistic than process oriented. Again, this is consistent with much of the historical effort in soil characterization and mapping. At the $i + 2$ and $i + 3$ scales, this remains true and appropriate today.

At the $i + 4$ level, attention shifts to areas defined by political boundaries, such as states, counties, and provinces. Multiple interacting watersheds, collected into units generally not defined by landscape features, present an additional modeling challenge. At this scale, as at the $i + 3$, there is decreased resolution in knowledge of individual soil properties, and in most cases only statistical models can be used. Manipulation of existing soil, climate, geohydrological and vegetation data bases must be accomplished by tools such as geographic information systems combined with spatial statistics. Conceptual and descriptive models are useful at these scales to generalize the geomorphic processes, and to relate landscape units. The influence of water and chemical fluxes on pedogenesis can be inferred, but not quantified, at this scale. Mapping by aerial photography identifies topographic and other important landscape features, such as fluvial depositions. Deterministic modeling at this scale is most often focused on management or regulatory issues, and in many cases models developed for use at smaller scales are used without regard to their limitations.

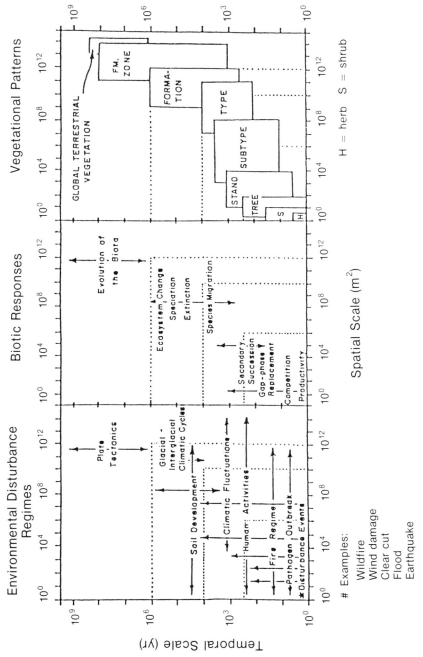

Fig. 2-2. Environmental disturbance regimes, biotic responses and vegetational patterns viewed in the context of four space-time domains (from Delcourt & Delcourt, 1988).

SCALE PARADIGM

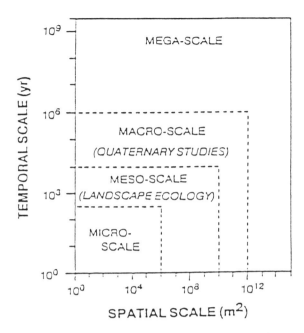

Fig. 2–3. Spatial-temporal domains for a hierarchial characterization of environmental forcing functions, biological responses and vegetational patterns (from Delcourt & Delcourt, 1988).

At the $i + 5$ (continental) and $i + 6$ (global) scales, conceptual models prevail. Soil orders and suborders (Soil Survey Staff, 1992) are the characterizations at this scale. Since all soils fit into 1 of only 10 orders, and there are then only 47 suborders for further distinction, pedogenetic processes must be broadly described. At this scale, the important issues of water and chemical fluxes related to pedogenesis are generally very broadly described, in terms of cause and effect, rather than mechanistic process. The presence or absence of diagnostic horizons that are indicative of parent material, biological activity and contrasting climates are the differentiating criteria, and are the basis of conceptual models of soil development in relation to landscape position, geohydrology, temperature and rainfall. As remote sensing techniques become more sophisticated further advances can be made at these scales in relating the consequences of man's activities to the continental scale impacts on pedogenesis (e.g., desertification, erosion).

Linkage between Levels

Clearly, each level of the hierarchy has a characteristic spatial and temporal scale, and it is not advisable to apply a model in a situation which does not match the scale for which it was originally developed. That is, it is insufficient to accomplish linkage between scales by taking a leaching model developed for the pedon scale and simply multiplying its result by some factor for the watershed scale. While it is *conceptually* true that process-level description at, for example, the pedon level could be used to extrapolate behavior at a watershed level, the uncertainties and difficulties of doing so should preclude the attempt. Spatial variation in properties and processes and the related spatial averaging that is needed for characterization, introduces such complexity and uncertainty that mechanistic supermodels are in the end probably no more accurate than the functional approaches described above. Recognizing this, a different conceptual approach should be taken that allows linkage of scales.

Simulation modeling at any *i*-level involves the application of accumulated knowledge at all lower *i*-levels, as represented in the immediately lower *i*-level. In such cases, it is usually not feasible to explicitly represent at the current level all those processes at the lower *i*-levels. Model complexity and data demands become overwhelming if this is attempted, and the objective of building mechanistically comprehensive supermodels is not what is needed. Rather, relevant processes at the lower *i*-levels should be considered as components of parameters defined in the next one or two upper levels. As an example, water and chemical fluxes at the lower *i*-levels are described by very basic forces that cause or result from molecular interactions. As the *i*-level increases, these forces are represented as an expression of soil water potential in mechanistic models, and at higher *i*-levels they are embodied in the changes in water content that are used to estimate water and chemical fluxes in functional models. This simple example is paralleled in other aspects of pedogenetic modeling, but is not often built on conceptually as modelers attempt to translate experimental evidence obtained at one scale into a model describing the next higher *i*-level.

It also is generally true that dynamic or mechanistic models applied at one level indicate the rates and directions of change at the next higher level. In pedogenesis, where processes extend over thousands of years, the lower level models can be used to provide periodic indications of the rates and directions of upper level processes. Subsequent alteration in the characteristics of the upper level such as changes in climate, hydrological patterns, biota, fire impose new boundary conditions that eventually necessitate re-evaluation of the magnitude, intensity and even nature of lower level processes. This re-evaluation then leads, perhaps, to different intensities and rates of weathering, erosion, accumulation and leaching. This alteration then is translated to higher levels, which may result in a new pedogenetic situation.

Mechanistic models provide information on rates of change and dynamics. They provide quantitative, although not necessarily accurate, predictions of processes. Knowledge of mechanistically described processes at scales one or two hierarchial levels lower is required. Functional models, on the other hand, are based on parameters that are easily estimated and can be defined within certain bounds (limits). They are useful to a greater extent in classification and rankings. Their output reflects the way in which processes and properties are classified, which is usually by easily measured parameters that have defined limits of variation. Thus the output of functional models may be constrained within certain limits, and are more useful for classification purposes than are mechanistic models.

The soil taxonomist interested only in classifying has generally chosen to use functional and conceptual models, but they only serve as tools limited by the extra assumptions that transform mechanistic rate models of water and chemical fluxes to functional models. The driving forces of water flow and chemical transport expressed in rate models allow pedologists of all sorts to cooperate in studies related to the understanding, integration, and (to a limited degree) extrapolation of these dynamic processes. Such cooperative efforts accomplished with full recognition of scale effects and appropriate modeling choices will increase our understanding of pedogenesis as a consequence of the driving forces of water and chemical fluxes.

SUMMARY

Pedogenesis depends on many environmental factors. Among these, the movement of water and the chemicals dissolved in it is a key process that influences the spatial distribution of some soil parent materials, but is ultimately more important to weathering and the formation of zones of eluviation and illuviation. Interest in soil behavior and management has created a consequent interest in the processes of pedogenesis, due to concerns over the ability to sustain the soil resource. Simulation modeling and improved instrumentation provide new tools in studies designed to provide insight into the underlying mechanisms of pedogenesis and man's impact on it. Recently, the application of hierarchial theory to pedogenesis has provided a conceptual framework within which models of water flow and chemical transport can be considered. This framework provides an opportunity for linkages to be established between spatial and temporal scales of natural processes. It also allows the separation of modeling efforts, and shows promise of helping to develop a new paradigm of the manner in which models of water and chemical fluxes are applied to different pedogenetic problems.

Major steps forward can now only be accomplished in this area through the cooperative efforts of several soil scientists including morphologists, chemists, mineralogists, biologists, and physicists. Such teams are difficult to assemble, and perhaps in the ideal also should be composed of other professionals in such areas as geology, landscape ecology, climatology, and engineering. Yet, if the science of pedogenesis is to advance, such teams must be organized, with their particular membership dependent on the scale of the modeling or experimental issues to be considered.

REFERENCES

Addiscott, T.M., and R.J. Wagenet. 1985. Concepts of solute leaching in soils: A review of modelling approaches. J. Soil Sci. 36:411–424.

Anderson, J.L., and J. Bouma. 1973. Relationship between hydraulic conductivity and morphometric data of an argillic horizon. Soil Sci. Soc. Am. Proc. 37:408–413.

Bootlink, H.W.G., and J. Bouma. 1991. Physical and morphological characterization of bypass flow in a well-structured clay soil. Soil Sci. Soc. Am. J. 55:1249–1254.

Bouma, J. 1989. Using soil survey data for quantitative land evaluation. p. 177–213. In B.A. Stewart (ed.) Advances in soil science. Vol. 9. Springer Verlag, New York.

Bouma, J., and J.L. Anderson. 1973. Relationships between soil structure characteristics and hydraulic conductivity. p. 77–105. In R.R. Bruce (ed.) Field soil moisture regime. SSSA Spec. Publ. 5. SSSA, Madison, WI.

Bouma, J., and J.L. Denning. 1974. A comparison of hydraulic conductivities calculated with morphometric and physical methods. Soil Sci. Soc. Am. Proc. 38:124–127.

Bouma, J., and H.A.J. van Lanen. 1987. Transfer functions and threshold values: From soil characteristics to land qualities. p. 106–111. *In* K.J. Beek et al. (ed.) Proc. ISSS/SSSA Workshop Quantified Land Evaluation Procedures, Washington, DC. 27 April to 2 May 1986. Int. Inst. Aerospace Surv. Earth Sci. Publ. no. 6. Enschede, the Netherlands.

Bouma, J., P.J.M. de Laat, R.H.C.M. Awater, H.C. van Heesen, A.F. van Holst, and Th.J. van de Nesl. 1980. Use of soil survey data in a model for simulating regional soil moisture regimes. Soil Sci. Soc. Am. J. 44:808–814.

Brewer, R. 1964. Fabric and mineral analysis of soils. John Wiley & Sons, Inc., New York.

Brimhall, G.H., O.A. Chadwick, C.J. Lewis, W. Compston, I.S. Williams, K.J. Danti, W.E. Dietrich, M.E. Power, D. Hendricks, and J. Bratt. 1991. Deformational mass transport and invasive processes in soil evolution. Science (Washington, DC) 255:695–702.

Chadwick, O.A., G.A. Brimhall, and D.M. Hendricks. 1990. From a black box to a gray box—a mass balance interpretation of pedogenesis. Geomorphology 3:369–390.

Delcourt, H.R., and P.A. Delcourt. 1988. Quaternary landscape ecology: Relevant scales in space and time. Landscape Ecol. 2:23–44.

Dijkerman, J.C. 1974. Pedology as a science: The role of data, models and theories in the study of natural soil systems. Geoderma 11:73–93.

Evans, C.V., and D.C. Roth. 1992. Conceptual and statistical modes to characterize soil materials, landforms and processes. Soil Sci. Soc. Am. J. 56:214–219.

Finke, P.A. 1993. Field scale variability of soil structure and its impact on crop growth and nitrate leaching in the analysis of fertilizing scenarios. Geoderma 60:89–107.

Havens, M.W. 1988. A GIS-based soil-landscape modeling approach to predict surface rock fragment distributions. M.S. thesis. The Pennsylvania State Univ.

Hinedi, Z.R., T.H. Skaggs, Z.J. Kabala, R.W.K. Lee, and A.C. Chang. 1992. Probing soil porosity with nuclear magnetic resonance. p. 238. *In* Agronomy abstracts. ASA, Madison, WI.

Hoogmoed, W.B., and J. Bouma. 1980. A simulation model for predicting infiltration into cracked clay soil. Soil Sci. Soc. Am. J. 44:458–461.

Hoosbeek, M.R., and R.B. Bryant. 1993. Towards the quantitative modeling of pedogenesis—A review. Geoderma 55:183–210.

Hutson, J.L., and R.J. Wagenet. 1992. LEACHM, leaching estimation and chemistry model. Version 3.0 Dep. SCAS Res. Rep. 92-3. Cornell Univ., Ithaca, NY.

Jenny, H. 1941. Factors of soil formation—A system of quantitative pedology. McGraw-Hill, New York.

Jenny, H. 1980. The soil resource: Origin and behavior. Ecol. Stad. 37. Springer Verlag, New York.

Johnson, D.L., E.A. Keller, and T.K. Rockwell. 1990. Dynamic pedogenesis: New views on some key soil concepts, and a model for interpreting Quaternary soils. Quaternary Res. 33:306–319.

Kirkby, M.J. 1985. A basis for soil profile modeling in a geomorphic context. J. Soil Sci. 36:97–121.

Lauren, J.G., R.J. Wagenet, J. Bouma, and H. Wosten. 1988. Variability of saturated hydraulic conductivity in a Glossaquic Hapludalf with macropores. Soil Sci. 145:20–28.

Lee, K-S., G.B. Lee, and E.J. Tyler. 1988. Thematic mapper and digital elevation modeling of soil characteristics in hilly terrain. Soil Sci. Soc. of Am. J. 52:1104–1107.

Levine, E.R., and E.J. Ciolkosz. 1986. A computer simulation model for soil genesis applications. Soil Sci. Soc. Am. J. 50:661–667.

Manley, E.P., W. Chesworth, and L.J. Evans. 1987. The solution chemistry of podzolic soils from the eastern Canadian shield: A thermodynamic interpretation of the mineral phases controlling soluble Al^{3+} and H_4SiO_4. J. Soil Sci. 38:39–51.

Marbut, C.F. 1928. A scheme for soil classification. p. 1–31. *In* Int. Soil Science, PROC. Congress. Vol. 4.

Marion, G.M., W.H. Schlesinger, and P.J. Fonteyn. 1985. CALDEP: A regional model for soil $CaCO_3$ (caliche) deposition in southwestern deserts. Soil Sci. 139:468–481.

Mayer, L. 1985. The distribution of calcium carbonate in soils: A computer simulation using program CALSOIL. U.S. Geol. Surv. 975. U.S. Geol. Surv., Menlo Park, CA.

Mizoguchi, M. 1992. Spectroscopic property of frozen clay-water system—Measurement of infrared spectra by FTIR. p. 223. *In* Agronomy abstracts. ASA, Madison, WI

Neustreuv, S.S. 1927. Genesis of soils. Russian Pedological Invest. III. Publ. Office Acad., Leningrad, USSR.

Nimah, M.N., and R.J. Hanks. 1973. Model for estimation of soil water, plant, and atmospheric interrelations: I. Description and Sensitivity. Soil Sci. Soc. Am. Proc. 37:522–527.

Petach, M.C., R.J. Wagenet, and S.D. DeGloria. 1991. Regional water flow and pesticide leaching using simulations with spatially-distributed data. Geoderma 48:245–270.

Peyton, R.L., B.A. Haeffner, S.H. Andersen, and C.J. Gantzer. 1992. Applying x-ray CT to measure macropore diameters in undisturbed soil cores. Geoderma 53:329–341.

Richardson, J.L., L.P. Wilding, and R.B. Daniels. 1992. Recharge and discharge of groundwater in aquic conditions illustrated with flow net analysis. Geoderma 53:65–78.

Shovic, H.F., and C. Montagne. 1985. Application of a statistical soil-landscape model to an order III wildland soil survey. Soil Sci. Soc. Am. J. 49:961–968.

Smeck, N.E., E.C.A. Runge, and E.E. Mackintosh. 1983. Dynamics and genetic modelling of soil systems. p. 23–49. *In* L.P. Wilding et al. (ed.) Pedogenesis and soil taxonomy 1. Concepts and interactions. Elsevier, Amsterdam.

Soil Survey Staff. 1992. Keys to soil taxonomy. 5th ed. Soil Manage. Support Serv. Tech. Monogr. 19. Pochontas Press, Inc., Blacksburg, VA.

Stumm, W., G. Furrer, E. Wieland, and B. Zinder. 1985. The effects of complex-forming ligands on the dissolution of oxides and aluminosilicates. p. 55–74. *In* J.I. Drever (ed.) The chemistry of weathering. Proc. NATO Advanced Res. Workshop on the Chemistry of Weathering, Rodez, France, NATO ASF Ser. C. Vol. 149.

Stein, A., I.G. Staritsky, J. Bouma, A.C. Van Eynsbergen, and A.K. Bregt. 1991. Simulation of moisture deficits and areal interpolation by universal co-kriging. Water Resour. Res. 27:1963–1973.

van Lanen, H.A.J., G.J. Reinds, O.H. Boersma, and J. Bouma. 1992. Impact of soil management systems on soil structure and physical properties in a clay loam soil and the simulated effects on water deficits, soil aeration and workability. Soil Tillage Res. 23:203–220.

Vepraskas, M.L., and J. Bouma. 1976. Model studies on mottle formation simulating field conditions. Geoderma 15:217–230.

Wagenet, R.J. 1983. Principles of salt movement in soil. p. 123–140. *In* D.W. Nelson et al. (ed.) Chemical mobility and reactivity in soil systems. SSSA Spec. Publ. 11. ASA and SSSA, Madison, WI.

Wagenet, R.J. 1990. Quantitative description of the leaching of organic and inorganic solutes in soil. p. 321–330. *In* D.J. Greenwood et al. (ed.) Quantitative theory in soil productivity and environmental pollution. Phil. Trans. R. Soc. London, B. 329:1255.

Wagenet, R.J., and P.S.C. Rao. 1990. Modeling pesticide fate in soil. p. 351–399. *In* H.H. Cheng et al. (ed.) Pesticides in the soil environment: Processes, impacts, and modeling. SSSA Book Ser. 2. SSSA, Madison, WI.

Wagenet, R.J., J. Bouma, and R.B. Grossman. 1991. Minimum data sets for use of soil survey information in soil interpretive models. p. 161–182. *In* M.J. Mausbach and L.P. Wilding (ed.) Spatial variabilities of soils and landforms. SSSA Spec. Publ. 28. SSSA, Madison, WI.

Wösten, J.H.M., C.H.J.E. Schuren, J. Bouma, and A. Stein. 1990. Comparing four methods to generate soil hydraulic functions in terms of their effect on simulated soil water budgets. Soil Sci. Soc. Am. J. 54:827–832.

3 Modeling Soil Solution, Mineral Formation and Weathering

Donald L. Suarez and Sabine Goldberg

USDA-ARS
Riverside, California

Simulation models are valuable tools to increase understanding of complex soil chemical processes. Initially modeling was restricted by limited computational power; thus, complex natural systems were often simulated with site-specific, regression-based models. These models (classified as functional models by Hoosbeek & Bryant, 1992) do not deal with the actual processes but were designed to simulate the response of the studied system to specific variables. Although useful for the conditions represented by the collected data set, these models are usually not suited for other environments. Often the process controlling variables or responses to these variables are different in different environments. Nonetheless, due to the complexity of natural systems, quantitative mechanistic models often do not exist.

The development of quantitative models has evolved from initial attempts to determine species distributions and ion activity, to models that attempt to represent chemical processes including soil mineral formation and prediction of soil solution composition and rates of mineral weathering. In this chapter we present an overview of the types of quantitative models suitable for describing soil chemical processes including mineral weathering. Generally following a chronological development, the models can be classified into a series of increasingly complex treatments of the natural system.

EQUILIBRIUM

Solution Speciation Models

The conceptual framework of solution speciation model was presented by Garrels and Christ (1965). One of the first chemical equilibrium models was IONPAIR developed by Thrailkill (1970). This relatively simple model required input of pH and alkalinity and computed the saturation status of a water with respect to calcite. The more extensive model WATEQ, developed by Truesdell and Jones (1974), included a complete speciation scheme for the major, naturally occurring chemical species and calculations of saturation status with respect to many important silicate minerals as well as some oxides and carbonates. This program has been updated periodically (WATEQ4F; Ball et al., 1987) and is still one of the most utilized models for speciation and saturation status of natural waters.

These speciation models are most useful for evaluating which soil processes are thermodynamically possible. By design they are not predictive models, as it is left to the user to evaluate which of the possible processes are dominant. This is in contrast to predictive models which force equilibrium with the most thermodynamically stable solid phases. This evaluation as to which process is dominant is clearly time dependent. Reactions that may be neglected for short time intervals may be important for longer time scales. By analyzing solution composition, soil genesis and possible weathering reactions can be considered based on the minerals present in the soil. The major disadvantage of these models for long-term soil studies is that computations, and subsequent predictions, are based on solution composition at the time of analysis. No processes are included to simulate temporal changes.

Despite the limitations of speciation programs, they usually serve as the foundation for more complex predictive models. A listing of some of the speciation models is presented in Table 3–1. Popular programs are GEOCHEM (Sposito & Mattigod, 1977) and MINTEQ (Felmy et al., 1984). These programs are mostly predictive programs, but they can be used for speciation alone by excluding solid phases. The major use of speciation models has been to input soil solution data obtained from soil which was previously collected in the field and reacted in the laboratory in a soil water suspension or saturation extract. In many instances, these conditions are not representative of the soil environments. Relatively few studies have used direct soil water analyses obtained either by soil water extractors or by squeezing water from freshly collected soil cores.

Predictive Models

Predictive models generally include detailed speciation routines but are intended to calculate the distribution of chemical species in the solution and solid phases. The models include a thermodynamic data base including solid phases and calculate concentrations based on thermodynamic equilibrium. The models are extensively used to predict solution composition in contact with soil and are excellent teaching tools. However, almost no studies have been conducted to compare these predictions to actual field conditions. Despite their limitations, these models are by far the most frequently used by soil scientists.

Thermodynamic equilibria models generally give good predictions for high temperature formation of mineral assemblages and rock-solution composition (>150°C). These models are less satisfactory for earth surface conditions, such as soils, where mixtures of mineral phases exist which are not in chemical equilibrium and where solution-mineral equilibrium rarely exists. Apparently reaction rates are too slow, at the relatively high water flux rates and relatively low (–10 to 40°C) temperatures to insure equilibria over short time scales (days–years). Nonetheless, the models can be used, as are the speciation models, to indicate which phases should be weathering or dissolving and which phases should be forming. This prediction of the final product is particularly useful when a speciation program calculates that a large number of phases are supersaturated. Output from the speciation program does not provide information about the relative

Table 3-1. Chemical models.

Solution speciation

Models	Sources	Attributes
IONPAIR	Thrailkill (1970)	Major ion activity calculations
SOLMNEQ	Kharaka & Barnes (1973)	Includes calculations of saturation status of soil minerals
WATEQ	Truesdell & Jones (1974)	Includes calculations of saturation status of soil minerals
WATEQ4F	Ball et al., 1987	Includes calculation of saturation status of soil minerals and various trace elements

Predictive

Models	Sources	Attributes
REDEQL2 GEOCHEM	McDuff & Morel (1973) Sposito & Mattigod (1977)	All can calculate saturation status or equilibrate solution with specified minerals
SOLMINEQ.88	Kharaka et al. (1988)	
SOILCHEM	Sposito & Coves (1988)	Can calculate saturation status or equilibrate solution with specified minerals. Contains constant capacitance model for adsorption
MINTEQ	Felmy et al. (1984)	Can calculate saturation status or equilibrate solution with specified minerals. Contains various surface complexation models for adsorption
HYDRAQL MICROQL FITEQL	Papelis et al., 1988 Westall, 1979 Westall, 1982	All contain constant capacitance, diffuse layer, stern and triple layer models

Table 3–1. Continued—Chemical models

| Reaction models | | | Transport models | | |
Models	Sources	Attributes	Models	Sources	Attributes
PHREEQE	Parkhurst et al., 1980	Provides equilibrium	HYDROGEOCHEM	Yeh & Tripath, 1990	Multicomponent transport with triple layer model. Possible to run unsaturated water flow
NETPATH	Plummer et al., 1991	Reaction path			
			DYNAMIX	Liu & Narasimhan, 1989	Multicomponent transport
			FASTCHEM	Hosteler & Erikson, 1989	Saturated water flow and multi-component transport
			TRANQL	Cederberg et al., 1985	Saturated water flow and multi-component transport with constant capacitance model
			SALT-FLOW SOWACH LEACHM NTRM	Robbins et al., 1980 Dudley & Hanks, 1991 Wagenet & Hutson, 1987 Schaffer & Larson, 1987	All include unsaturated water flow, multicomponent transport, ion exchange, and calcite and gypsum equilibiria
			SOILCO2	Simunek & Suarez, 1993	Unsaturated water flow, CO_2 production and transport
			UNSATCHEM	Suarez & Simunek, 1992	Unsaturated water flow, multi-component transport, ion exchange, calcite and dolomite kinetics, gypsum equilibrium, CO_2 production and transport, silicate weathering kinetics

stabilities of the mineral phases or the path or evolution of the system. The predictive models are particularly useful for soil genesis problems, but do not consider intermediate unstable phases that may be transitional but nonetheless very significant. The models appear most useful for conceptually demonstrating how a solution or solid is altered at equilibrium (for example where framework silicates such as feldspars weather to clays and oxides), but are not as useful for prediction of the solution or solid composition. Use of the models for realistic simulations rather than thermodynamic calculations requires that at least some kinetic aspects be considered and the thermodynamic equilibrium criteria be relaxed. For example, if quartz is omitted from the data base, predicted silica concentrations increase and clay formation can be predicted.

As expected, the models differ in their input requirements, reactions considered, and ease of use. The model GEOCHEM (Sposito & Mattigod, 1977), and SOILCHEM (Sposito & Coves, 1988) require input of total inorganic C and pH or CO_2 partial pressure. This information is not usually available from soil water analysis but can be estimated by iteration by the user if only alkalinity (and pH or pCO_2) is available. Conservation of total inorganic C is not a useful model for soil reactions (field) since soils are open to the atmosphere and biological production and transport of CO_2 are important aspects of the soil C cycle. These models consider a large number of solid phases including trace metal minerals that may not exist in soils due to kinetic constraints.

Among other commonly used predictive programs is MINTEQ (Felmy et al., 1984), initially based on the speciation program WATEQ (Truesdell & Jones, 1974). The model MINTEQ accepts either total soluble inorganic C or alkalinity in the input file. Most water analyses report alkalinity, but dissolved inorganic C is rarely available. A listing of several widely used thermodynamic equilibrium programs that consider solid phases is presented in Table 3–1.

Recent chemical models have related reaction rates to concentrations of adsorbed species rather than concentrations in the bulk solution. The following sections examine adsorption modeling, which is required for modeling most mineral weathering rates, as well as for ion adsorption.

ADSORPTION MODELS

Adsorption is the net accumulation of a substance at the interface between a solid phase and an aqueous solution phase (Sposito, 1989). Numerous models exist to describe adsorption reactions. Empirical models have been utilized most often to provide descriptions of adsorption data but are limited by their lack of a theoretical basis. These models are special cases of the generalized empirical adsorption isotherm equation

$$x = \frac{bKc^\beta}{1 + Kc^\beta}$$ [1]

where x is the amount adsorbed per unit mass, c is the equilibrium solution concentration, and b, K, and β are empirical parameters (Goldberg & Sposito, 1984).

The Langmuir adsorption isotherm equation was developed to describe the adsorption of gases on clean surfaces, but nevertheless has been used often to describe ion adsorption by soils and soil minerals. The Langmuir equation is a special case of Eq. [1] where $\beta = 1$ (Goldberg & Sposito, 1984)

$$x = \frac{bKc}{1 + Kc}$$

[2]

For many studies the Langmuir equation can describe ion adsorption only under conditions of low solution concentration. The Langmuir equation implies uniform surface sites and absence of lateral interactions.

The Freundlich adsorption isotherm equation is strictly valid only for adsorption at low concentrations (Sposito, 1984) but has often been used to describe ion adsorption by soils and soil constituents over the entire concentration range studied. The Freundlich equation is a special case of Eq. [1] where $b = 1$, $0 < \beta < 1$, $Kc^\beta < < 1$ (Goldberg & Sposito, 1984)

$$x = Kc^\beta$$

[3]

The Freundlich equation implies heterogeneity of surface sites. The Freundlich equation is often used empirically in situations where $Kc^\beta > > 1$.

Although the Langmuir and Freundlich equations are often excellent at describing ion adsorption, they are numerical relationships used to curve-fit data (Harter & Smith, 1981). Independent experimental evidence for adsorption must be present before any chemical meaning can be assigned to Langmuir and Freundlich equation parameters. Since the use of the Langmuir and Freundlich equations is essentially a curve-fitting procedure, the parameters are valid only for the conditions under which the experiment was conducted. The models do not consider changes in adsorption as a function of pH or other chemical or physical factors.

Unlike empirical models, surface complexation models are chemical models that provide a general molecular description of adsorption phenomena using an equilibrium approach. The surface complexation models are designed to calculate values of thermodynamic properties such as activity coefficients and equilibrium constants mathematically. The most significant advancement in these models is that they consider the charge on both the adsorbing ion and the adsorbent surface. Five common surface complexation models are: the constant capacitance model (Stumm et al., 1980), the triple layer model (Davis et al., 1978), the Stern variable-charge, variable-surface potential model (Bowden et al., 1980), the generalized two-layer model (Dzombak & Morel, 1990), and the one-pK model (van Riemsdijk et al., 1986). A detailed discussion of these models is presented elsewhere (Goldberg, 1992). All models contain the following adjustable parameters: K_ι, the surface complexation constants; C_ι, the capacitance density for the ιth surface plane; and $[SOH]_T$, the total number of reactive surface hydroxyl groups. A summary of the more frequently used constant capacitance model and triple layer model will be provided here.

The constant capacitance model was developed by Schindler, Stumm, and their coworkers (Schindler & Gamsjäger, 1972; Hohl & Stumm, 1976; Schindler et al., 1976; Stumm et al., 1976, 1980). The model is based on the following assumptions: ion adsorption is based on a ligand exchange mechanism; all surface complexes are inner-sphere complexes; no surface complexes are formed with ions from the background electrolyte; the relationship between surface charge and surface potential is

$$\sigma = \frac{CSa}{F}\psi$$

[4]

where C is the capacitance (F m^{-2}), S is the surface area (m^2 g^{-1}), a is the suspension density (g L^{-1}), F is the Faraday constant (C mol^{-1}), and σ has units mole charge per liter. A diagram of the surface-solution interface for the constant capacitance model is provided in Fig. 3–1. Applications to natural systems are complicated by the requirement of constant ionic strength.

The equations for the inner-sphere surface complexation reactions are (Hohl et al., 1980)

$$SOH + H^+ \Leftrightarrow SOH_2^+$$

[5]

$$SOH \Leftrightarrow SO^- + H^+$$

[6]

$$SOH + M^{m+} \Leftrightarrow SOM^{(m-1)} + H^+$$

[7]

$$2SOH + M^{m+} \Leftrightarrow (SO)_2 M^{(m-2)} + 2H^+$$

[8]

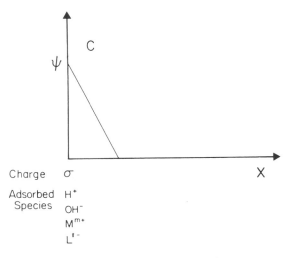

Fig. 3–1. Placement of ions, potential, charge, and capacitance for the constant capacitance model (after Westall, 1986).

$$SOH + L^{\ell^-} \Leftrightarrow SL^{(\ell-1)^-} + OH^-$$ [9]

$$2SOH + L^{\ell^-} \Leftrightarrow S_2L^{(\ell-2)^-} + 2OH^-$$ [10]

where *SOH* represents the surface functional group, *M* is a metal ion, m^+ is the charge on the metal ion, *L* is a ligand, and ℓ^- is the charge on the ligand.

The intrinsic conditional equilibrium constants describing the above reactions are (Hohl et al., 1980)

$$K_+(int) = \frac{\left[SOH_2^+\right]}{\left[SOH\right]\left[H^+\right]} \exp\left[F\psi / RT\right]$$ [11]

$$K_-(int) = \frac{\left[SO^-\right]\left[H^+\right]}{\left[SOH\right]} \exp\left[-F\psi / RT\right]$$ [12]

$$K_M^1(int) = \frac{\left[SOM^{(m-1)}\right]\left[H^+\right]}{\left[SOH\right]\left[M^{m+}\right]} \exp\left[(m-1)F\psi / RT\right]$$ [13]

$$K_M^2(int) = \frac{\left[(SO)_2 M^{(m-2)}\right]\left[H^+\right]^2}{\left[SOH\right]^2\left[M^{m+}\right]} \exp\left[(m-2)F\psi / RT\right]$$ [14]

$$K_L^1(int) = \frac{\left[SL^{(\ell-1)^-}\right]\left[OH^-\right]}{\left[SOH\right]\left[L^{\ell-}\right]} \exp\left[-(\ell-1)F\psi / RT\right]$$ [15]

$$K_L^2(int) = \frac{\left[S_2L^{(\ell-2)^-}\right]\left[OH^-\right]^2}{\left[SOH\right]^2\left[L^{\ell-}\right]} \exp\left[-(\ell-2)F\psi / RT\right]$$ [16]

where *R* is the molar gas constant (J mol^{-1} K^{-1}), *T* is the absolute temperature (K), and square brackets represent concentrations (mol L^{-1}). Values of the intrinsic conditional equilibrium constants can be obtained by extrapolating the conditional equilibrium constants to zero net surface charge (Stumm et al., 1980). A detailed explanation of the procedure is provided in Stumm et al. (1980). An example of the ability of the constant capacitance model to describe adsorption data is indicated in Fig. 3–2 for silicate adsorption on an amorphous aluminum oxide.

The original triple layer model was developed by Davis and coworkers (Davis et al., 1978; Davis & Leckie, 1978, 1980) as an extension of the site binding model (Yates et al., 1974). The model is based on the following assumptions: only protons and hydroxyl ions form inner-sphere complexes; all ion adsorption reactions form outer-sphere complexes; three planes of charge represent the surface; the relationships between surface charges and surface potentials are

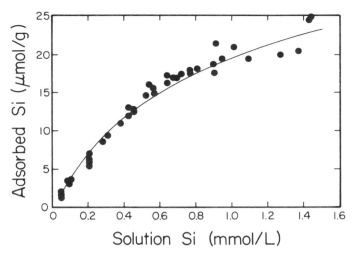

Fig. 3–2.   Silicate adsorption on amorphous aluminum oxide at pH 9.2 with experimental data (circles) from Hingston and Raupach (1967). The solid line represents model results: $\log K^1_{Si}(int)$ = 3.29, $\log K^2_{Si}(int)$ = –7.93.

$$\psi_0 - \psi_\beta = \sigma_0 / C_1 \qquad [17]$$

$$\psi_\beta - \psi_d = -\sigma_d / C_2 \qquad [18]$$

$$\sigma_d = -\left(8RTc\varepsilon_0 D\right)^{1/2} \sinh\left(F\psi_d / 2RT\right) \qquad [19]$$

where ε_0 is the permittivity of a vacuum, D is the dielectric constant of water, c is the concentration of a 1:1 background electrolyte, and σ has units of C per meters squared. The triple layer model has been modified to include ion adsorption as inner-sphere surface complexes (Blesa et al., 1984; Hayes & Leckie, 1986). A diagram of the surface-solution interface for the triple layer model is provided in Fig. 3–3.

The equations for the inner-sphere surface complexation reactions in the triple layer model are Eq. [5] through [10] as written for the constant capacitance model. The equations for the outer-sphere surface complexation reactions are (Davis et al., 1978; Davis & Leckie, 1978, 1980):

$$SOH + M^{m+} \Leftrightarrow SO^- - M^{m+} + H^+ \qquad [20]$$

$$SOH + M^{m+} + H_2O \Leftrightarrow SO^- - MOH^{(m-1)} + 2H^+ \qquad [21]$$

$$SOH + H^+ + L^{\ell-} \Leftrightarrow SOH_2^+ - L^{\ell-} \qquad [22]$$

$$SOH + 2H^+ + L^{\ell-} \Leftrightarrow SOH_2^+ - LH^{(\ell-1)-} \qquad [23]$$

$$SOH + C^+ \Leftrightarrow SO^- - C^+ + H^+ \qquad [24]$$

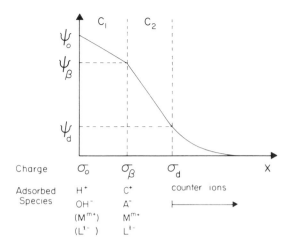

Fig. 3–3. Placement of ions, potentials, charges, and capacitances for the triple layer model. Parentheses represent ion placement allowed only in the modified triple layer model (after Westall, 1980).

$$SOH + H^+ + A^- \Leftrightarrow SOH_2^+ - A^-$$ [25]

where C^+ is the cation and A^- is the anion of the background electrolyte.

The intrinsic conditional equilibrium constants for the triple layer model are Equations [11] through [16] as written for the constant capacitance model for inner-sphere surface complexes. For outer-sphere surface complexes the intrinsic conditional equilibrium constants are given in Davis et al., 1978; Davis & Leckie, 1978, 1980. As for the constant capacitance model, values for the intrinsic conditional equilibrium constants can be obtained by extrapolation. A detailed explanation of various types of extrapolation procedures used for the triple layer model is provided in Goldberg (1992). The model has the potential to be used under conditions of varying ionic strength as it considers changes in surface potential with changing electrolyte concentration. Figure 3–4 provides an example of the ability of the triple layer model to describe adsorption data for Cu adsorption on amorphous iron oxide.

Adsorption models have been incorporated into various chemical speciation models. The computer program MINTEQ (Felmy et al., 1984) combines the thermodynamic data base of WATEQ3 (Ball et al., 1981) into the mathematical framework of the computer program MINEQL (Westall et al., 1976) and contains surface complexation models. The program MINTEQA1 (Brown & Allison, 1987) contains the Langmuir and Freundlich equations, the constant capacitance model and the triple layer model. The model MINTEQA2 (Allison et al., 1990) has also added the diffuse layer model. The chemical speciation program SOILCHEM (Sposito & Coves, 1988) contains the constant capacitance model. The computer programs HYDRAQL (Papelis et al., 1988), MICROQL (Westall, 1979), and FITEQL (Westall, 1982) all contain the constant capacitance model, the diffuse layer model, the Stern model, and the triple layer model.

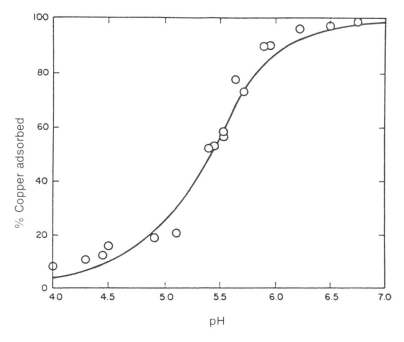

Fig. 3–4. Copper adsorption on amorphous iron oxide. Experimental data-circles. Solid line represents model results: $\log K^1_{Cu}(\text{int}) = -4.1$, $\log K^2_{Cu}(\text{int}) = -9.0$, (from Davis & Leckie, 1978).

EFFECTS OF SURFACE COMPLEXATION REACTIONS ON THE KINETICS OF DISSOLUTION

The dissolution kinetics of most silicates and oxides appears to be controlled by chemical surface processes (Furrer & Stumm, 1986, Stumm & Wollast, 1990), rather than transport through a surface layer. As a result it is necessary that the concentration of surface species rather than solution concentrations be considered for expressing reaction rates. The reaction steps are: attachment of the reactants at surface sites; rate-limiting detachment of the surface metal species, transport of the metal complex into the bulk solution, and regeneration of the active site by protonation. Under acid conditions the dissolution is promoted by protons that bind to the surface oxide ions (Furrer & Stumm, 1986). Calculation of surface species concentrations requires that activities rather than concentrations be used for the relevant species in solution. Figure 3–5 is a schematic of the proton-promoted dissolution of a trivalent oxide. The proton-promoted dissolution rate is (Stumm & Wollast, 1990)

$$R_H = k_H \left[SOH_2^+ \right]^n = k_H \left(C_H^S \right)^n \tag{26}$$

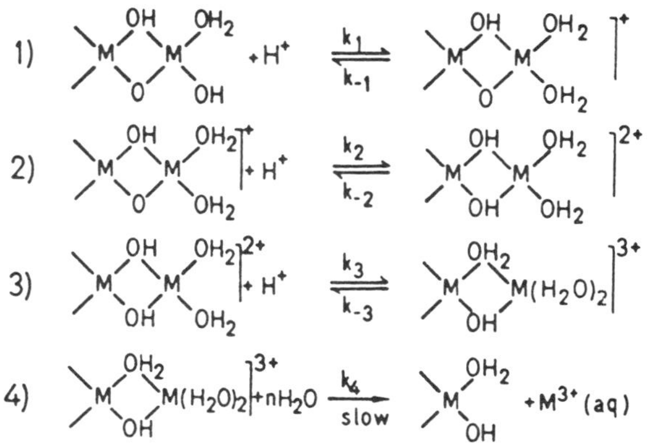

1) $\quad \cdots + H^+ \underset{k_{-1}}{\overset{k_1}{\rightleftharpoons}} \cdots$

2) $\quad \cdots + H^+ \underset{k_{-2}}{\overset{k_2}{\rightleftharpoons}} \cdots$

3) $\quad \cdots + H^+ \underset{k_{-3}}{\overset{k_3}{\rightleftharpoons}} \cdots$

4) $\quad \cdots + nH_2O \underset{slow}{\overset{k_4}{\longrightarrow}} \cdots + M^{3+}(aq)$

Fig. 3–5. Proton-promoted dissolution of a trivalent oxide M_2O_3 (from Furrer & Stumm, 1986).

where k_H is the rate constant, C^s_H is the surface proton concentration in moles per meter squared, and n is the number of protonation steps in the dissolution mechanism. The concentration of protonated surface hydroxyl groups, $[SOH_2^+]$ can be obtained using the constant capacitance model or determined from titration data. In the presence of complex-forming ligands the dissolution is promoted by ligands that form surface complexes by ligand exchange (Furrer & Stumm, 1986). Figure 3–6 shows the ligand-promoted dissolution of a trivalent oxide. The ligand-promoted dissolution rate is (Stumm & Wollast, 1990)

$$R_L = k_L[SL] = k_L C^S_L \qquad [27]$$

where k_L is the rate constant and C^s_L is the concentration of adsorbed ligand in moles per meter squared. The concentration of surface sites occupied by ligand, [SL] is obtained by adsorption studies and can be modeled using the constant capacitance model. The proton-promoted and the ligand-promoted dissolution are considered independent and the total rate is additive (Furrer & Stumm, 1986)

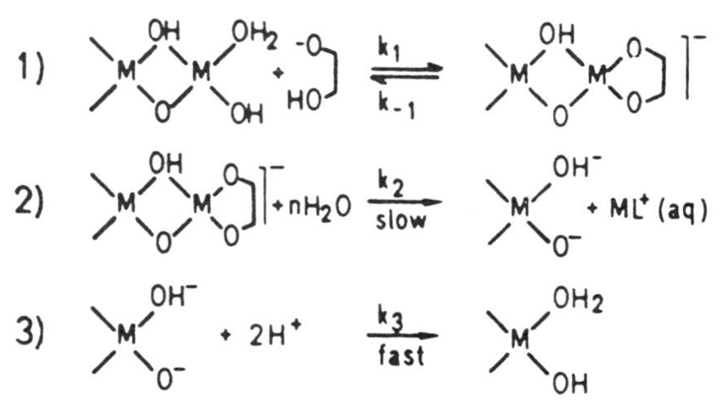

1) $\quad \cdots \underset{k_{-1}}{\overset{k_1}{\rightleftharpoons}} \cdots$

2) $\quad \cdots + nH_2O \underset{slow}{\overset{k_2}{\longrightarrow}} \cdots + ML^+(aq)$

3) $\quad \cdots + 2H^+ \underset{fast}{\overset{k_3}{\longrightarrow}} \cdots$

Fig. 3–6. Ligand-promoted dissolution of a trivalent oxide M_2O_3 (from Furrer & Stumm, 1986).

$$R_{tot} = R_H + R_L \qquad\qquad [28]$$

Figure 3–7 is a schematic of the ligand- and proton-promoted dissolution of a trivalent oxide. The dependence of the rate of the proton-promoted dissolution of δ-Al_2O_3 and α-FeOOH on the surface concentration of protons is indicated in Figure 3–8. The slopes are close to three indicating that n is equal to three for δ-Al_2O_3 and α-FeOOH. Furrer and Stumm (1986) suggested that the reaction order n may correspond to the oxidation number of the metal ion in the oxide. Figure 3–9 indicates the dependence of the rate of the ligand-promoted dissolution of δ-Al_2O_3 on the surface concentration of organic ligands.

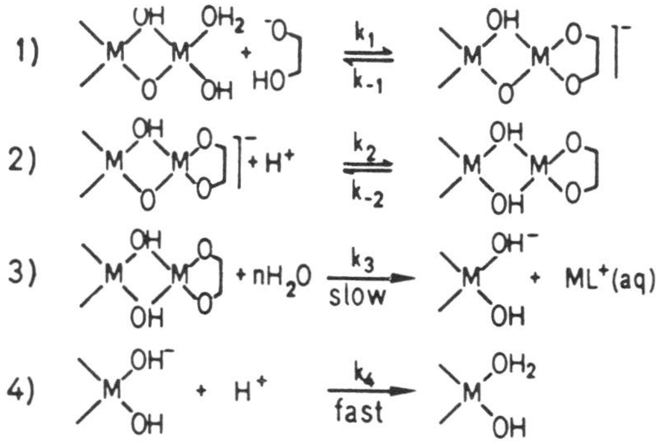

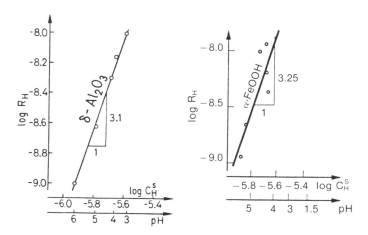

Fig. 3–7. Ligand/proton-promoted dissolution of a trivalent oxide M_2O_3 (from Furrer & Stumm, 1986).

Fig. 3–8. Dependence of the rate of the proton-promoted dissolution of δ-Al_2O_3 and α-FeOOH (from Stumm & Wollast, 1990, based on data from Furrer & Stumm, 1986, for δ-Al_2O_3 and Zinder et al., 1986 for α-FeOOH).

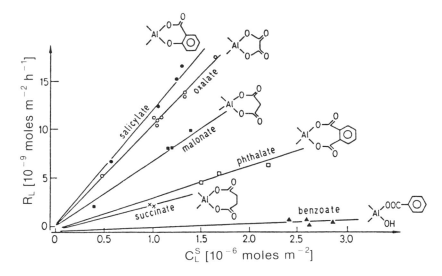

Fig. 3–9.   Dependence of the rate of the ligand-promoted dissolution of δ-Al$_2$O$_3$ on the surface concentration of ligands (from Stumm et al., 1987, based on data from Furrer & Stumm, 1986).

The Furrer and Stumm (1986) dissolution model requires modification to represent silicate dissolution rates. Helgeson et al. (1984) did not express their data in terms of adsorbed proton concentrations but reported that feldspar dissolution rates were independent of pH in the range of pH 2.9 to 8.0. Anorthite dissolution rates shown in Fig. 3–10 (Amrhein & Suarez, 1988) indicate that feldspar dissolution cannot be represented directly with the oxide model. Amrhein and Suarez (1988) modified the Furrer and Stumm model by adding a

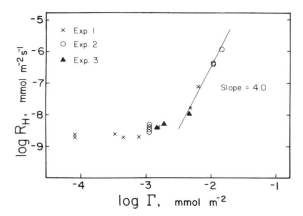

Fig. 3–10.   The log of the proton-promoted rate vs. the log of the net concentration of adsorbed protons and hydroxide ions (Amrhein & Suarez, 1988).

rate term proportional to the quantity of uncharged silanol sites (SOH). In addition, the dissolution rate at high pH was represented by the adsorbed hydroxyl concentration, whose reaction rate was equal to the proton promoted rate. Thus the term C_H^s was replaced by the term Γ, which then represented the sum of the adsorbed proton and hydroxyl concentrations. Although there is evidence that the order of the proton and hydroxyl promoted rates are similar (Furrer & Stumm, 1986) this result does not mean that the reaction rate constant must be the same. Schott (1990) determined that for the dissolution of a basalt glass, the reaction order was 3.8 for protons and 3.7 for hydroxyls, but the reaction rate for hydroxyls was about seven times greater than for protons.

The anorthite reaction rate in the presence of organic ligands was related to the adsorbed ligand concentration, as in the Furrer and Stumm model for oxides. Figure 3–11 shows the ligand-promoted dissolution rate for anorthite (Amrhein & Suarez, 1988). Combining these terms produces an overall rate expression for anorthite (Amrhein & Suarez, 1988)

$$R_T = 37.0(\Gamma)^{4.0} + 2.09 \times 10^{-8}(\text{SOH}) + 4.73 \times 10^{-6}(\text{S}-\text{L}). \qquad [29]$$

Differences in reaction dependence on ligand concentration are also related to the Al content and likely to the Fe content. The rate term k_L for the plagioclase feldspar series is proportional to the Al content, as shown in Fig. 3–12 (Amrhein & Suarez, 1988). This concept can likely be generalized to most minerals, as Bales and Morgan (1986), and Bennett et al., (1988) indicated no effect of ligands on chrysotile and quartz dissolution, respectively. These organic ligands do not form important complexes with silica. It appears likely that the ligand promoted rate is dominant under the pH and ligand concentration range of natural systems.

Application of these reaction models to predict reaction rates in natural systems is not yet possible. Dissolution rates are related to the concentration of active sites, which are related indirectly to the measurable entity, surface area.

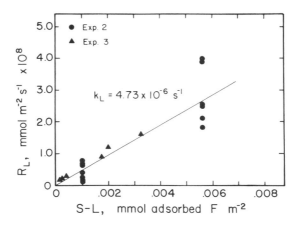

Fig. 3–11. Ligand-promoted dissolution rate of anorthite vs. the concentration of adsorbed F⁻ (Amrhein & Suarez, 1988).

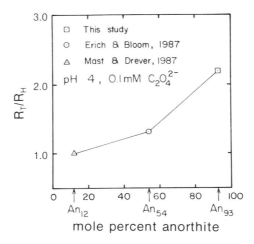

Fig. 3–12. Effect of oxalate on the anorthite dissolution rate vs. the mole percent Ca in the plagioclase. A ratio of 1.0 represents no effect of oxalate on the rate of dissolution (Amrhein & Suarez,
1988).

Laboratory reaction rates have been demonstrated to be orders of magnitude
faster than measured field dissolution rates. This result is related both to the number of active sites per meter squared surface area, as well as to the adsorption of
inhibiting ions. Reduction of reaction rates with time may be related to release of
Al and possible absorption of Al onto the surface silanol polymers causing dehy-
dration and crosslinking (Casey et al., 1988). Evidence for this mechanism is provided by the fact that above pH = 3, Al addition has been observed to decrease
feldspar dissolution rates (Amrhein & Suarez, 1992). Addition of K^+ to solution
also causes a reduction in the dissolution rate (Amrhein & Suarez, 1992), apparently due to the competition of K^+ for protons in the cavities of the siloxane surface. True back reaction effects have not been observed, probably because these
aluminosilicates are far from equilibrium in these experiments as well as in
almost all natural systems in earth surface environments. Long-term reaction
rates (after 4.5 yr) of specimen minerals in the laboratory decreased to below
0.005 times the initial rates (Amrhein & Suarez, 1992). These long-term rates are
similar to feldspar reaction rates determined in the laboratory for material separated from soils (Suarez & Wood, 1991) as well as reaction rates calculated under
field conditions (Velbel, 1986).

Carbonate Kinetics

Detailed carbonate reaction models have been developed by Plummer et al.
(1978) for calcite and Busenberg and Plummer (1982) for dolomite. These
authors determined that the dissolution process was controlled by different rate
limiting steps, dependent on pH. The overall reaction expression is

$$R = k_1 \left(H^+ \right) = k_2 \left(H_2CO_3^* \right) + k_3 \left(H_2O \right) - k_4 \left(Ca^{2+} \right) \left(HCO_3^- \right) \qquad [30]$$

where

$$k_4 = \frac{K_2}{K_c}\left(k_1 + \frac{1}{\left(H_s^+\right)}\left[k_2\left(H_2CO_3^*\right)_s + k_3\left(H_2O\right)_s\right]\right)$$

[31]

and k_1, k_2 and k_3 are first-order rate constants dependent on temperature; k_4 is the backward rate constant; K_2 is the second dissociation constant for carbonic acid; K_c is the solubility product of calcite, and $H_2CO_3^*$ is the sum of undissociated carbonic acid and dissolved aqueous CO_2. The value H_s^+ represents the H activity at the calcite surface. Similarly the dolomite rate expression of Busenberg and Plummer (1982) below 45°C is

$$R = k_1\left(H^+\right)^{0.5} = k_2\left(H_2CO_3\right)^{0.5} + k_3\left(H_2O\right)^{0.5} - k_4\left(HCO_3^-\right)$$

[32]

where k_1, k_2 and k_3 again are the forward rate constants and k_4 is the backward rate constant. Suarez (1983), Inskeep and Bloom (1985), and Chou et al. (1989) in detailed experiments, observed that the Plummer et al. (1978) model underestimated precipitation rates at low CO_2. A simpler expression for precipitation is given by Inskeep and Bloom (1985) who modified the Nancollas and Reddy (1971) precipitation model to account for changes in ionic strength. The proposed precipitation model is

$$R = \gamma_2^2 k_f S\left[Ca^{2+}\right]\left[CO_3^{2-}\right] - K_c\gamma_2^{-2}$$

[33]

where k_f = 118 L^2 mol^{-1} m^{-2} s^{-1} γ is the divalent ion activity coefficient and S is the calcite surface area in meters squared per liter (Inskeep & Bloom, 1985). Chou et al. (1989) examined carbonate precipitation and dissolution under low CO_2 conditions. They developed the relationship

$$R = k_1\left(H^+\right) + k_2\left(H_2CO_3^*\right) + k_3\left(H_2O\right) - k_{-3}\left(Ca^{2+}\right)\left(CO_3^{2-}\right)$$

[34]

where log k_1, log k_2, log k_3 equal –4.05, –7.30, –10.19 respectively, and log k_{-3}, equals –1.73 for calcite. The back reaction value k_{-3} is very close to that determined by Inskeep and Bloom (1985).

Use of these crystal growth models provides a description of the changes in ion concentrations with time but may not be mechanistically applicable for soil systems. Inskeep and Bloom (1986) determined that dissolved organic carbon (DOC) values of 146 μmol L^{-1} resulted in complete inhibition of calcite crystal growth. Suarez et al. (1992) measured DOC values of 500 to 2000 μmol L^{-1} in suspensions of arid land soils that were supersaturated and not precipitating calcium carbonate. From these data it seems that crystal growth is not the major process controlling calcite precipitation and that heterogeneous nucleation may be more realistic. Recently, Svensson and Dreybrodt (1992) modified the Plummer et al. (1978) model to account for reduction in precipitation rates (crystal growth) due to adsorption of inhibitors. The data show that the effect of the inhibitors is not constant, but rather increases dramatically as the solution approaches calcite equilibrium.

Transport Models

To provide a more realistic representation of natural systems, it is necessary to consider water flow and the spatial differences in chemical and mineralogical properties of the soil. The following sections describe some modeling approaches useful to soil systems.

Reaction Path

Reaction path models provide a more realistic representation of chemical reactions in soils than do the batch equilibrium models. Reaction path models have not been generally utilized in soil science but represent a substantial advance over the batch reaction models. Simulations with these models have been made in hydrochemical and geochemical studies both to evaluate evolution of groundwaters along flowpaths in aquifers, as well as, to calculate necessary water composition and quantify boundary conditions for specified rock alterations. Among the available models, PHREEQE (Parkhurst et al., 1980) and NETPATH (Plummer et al., 1991) are likely the most utilized, and are based on the WATEQ speciation model and data base. Other available models are listed in Table 3–1.

These reaction path models contain mass balance relations for both the solid and solution phases, thus they can be used to simulate soil weathering processes. Although not yet extensively used for pedogenic studies, they are well suited for this purpose.

Multicomponent Transport

Saturated and/or Steady-State Water Flow

The unnamed model of Jury et al. (1978), couples steady state water flow with a chemical equilibrium model. The model considers major ion chemistry, ion exchange, and the possibility to dissolve or precipitate calcite or gypsum to equilibrium. Water uptake is accounted for by specifying the leaching fraction and proportioning the water uptake at various depths in the root zone. The model is suited for generation of calcite and gypsum distribution profiles. Other models in this category include the model of Schulz and Reardon (1983) and the model CALDEP (Marion, et al., 1985). The model CALDEP, discussed in detail in Marion and Schlesinger (see Chapter 8; 1994) includes a "tipping bucket" flow model (water flow from one layer to another only when saturation is achieved) and a simplified chemical model to enable long-term simulations.

Other more generalized multicomponent transport models include HYDRO-GEOCHEM (Yeh & Tripathi, 1989), DYNAMIX (Liu & Narasimhan, 1989), and FASTCHEM (Hostetler & Erikson, 1989). The model HYDROGEOCHEM has the added advantage that it also can be used for unsaturated water flow by running a separate water flow program and inputing the results to HYDRO-GEOCHEM. These generalized programs are very flexible in their ability to handle different chemical environments but do require the user to define chemical species, solids, and thermodynamic constants for the reactions. The models can consider reduction-oxidation dependent species, carbonates and silicates. All models require input of pH or pCO_2. This requirement prevents them from being used for long-term simulations, such as long-term acidification where these variables change.

Surface complexation models have not yet been incorporated widely into transport models (Mangold & Tsang, 1991). The computer program TRANQL (Cederberg et al., 1985) and an unnamed program by Jennings et al. (1982) have incorporated the constant capacitance model into transport programs. The transport model HYDROGEOCHEM(Yeh & Tripathi, 1990) contains the triple layer model.

Unsaturated Water Flow

These models have all been developed primarily for soil environments in general and are oriented to agricultural lands. Among the fixed pH models are SALT-FLOW (Robbins et al., 1980), LEACHM I (Wagenet & Hutson, 1987), and NTRM (Schaffer & Larson, 1987), and SOWACH (Dudley & Hanks, 1991). These models are all primarily major species models that consider ion exchange and calcite and gypsum precipitation-dissolution, although some (e.g., LEACHM I and NTRM) also include N species. These models are primarily useful for intermediate-term simulations where pH is not changing over the long term and where transient changes can be neglected. The model LEACHM II (discussed in detail by Wagenet et al., 1994; see Chapter 2) considers fixed CO_2 as an input, thus allowing for changes in soil pH with time associated with changes in input water chemistry.

Chemical Kinetic Models

The model UNSATCHEM (Suarez & Simunek, 1992) combines the program SOILCO2 (Simunek & Suarez, 1993) with an unsaturated water flow-major ion chemistry program. The model differs from other soil simulation models in that the CO_2 content is calculated based on environmental factors (water inputs, temperature, soil properties), rather than used as an input. Prediction of pH is made by combining the CO_2 distributions with the chemistry routine, requiring only the input of major ion solution composition (rainfall or irrigation). The model includes a chemical speciation routine, cation exchange and mineral reactions. Use of kinetic expressions in UNSATCHEM for calcite dissolution-precipitation and dolomite dissolution allows for more realistic representations of soil conditions than simulations utilizing mineral equilibrium conditions.

Figure 3–13 presents the predicted calcite distribution with depth at various times for a 1-yr cycle of irrigation using lower Colorado River water. This simulation using the equilibrium option in the UNSATCHEM model (Suarez & Simunek, 1992) assumes a constant water application of 1 cm d^{-1} and a water extraction of 0.09 cm d^{-1}. The simulation assumes a fixed atmospheric CO_2 upper boundary and a linearly increasing CO_2 concentration reaching a value of 2 kPa. The water content distribution stabilized after about 200 d (data not shown). Since Colorado River water is initially calcite supersaturated, large quantities of calcite are precipitated in the first node (Fig. 3–13a). Below the first node, increasing CO_2 results in calcite dissolution and then subsequent precipitation. Eventually the steady-state profile was reached after 200 d, showing a sharp precipitation front at 55 cm. The same simulation using the kinetic option in

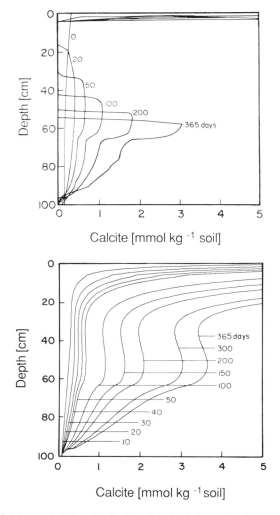

Fig. 3–13. (a) Calcite precipitation distribution with depth for a 1-yr irrigation sequence using Colorado River water and the UNSATCHEM model (Suarez & Simunek, 1992, p. 32, in preparation) with the equilibrium option; (b) Calcite precipitation distribution with depth for a 1-yr irrigation sequence using Colorado River water and the UNSATACHEM model (Suarez & Simunek, unpublished data) with the nonequilibrium option.

UNSATCHEM presents a much different calcite distribution, as shown in Fig. 3–13b. In addition to the differences in the calcite distribution, the increased concentrations of Ca in solution with the kinetic option result in gypsum precipitation at a shallower depth as compared with the results with the equilibrium option. These differences have obvious implications both for predicting soil mineral weathering and dissolution and for reconstruction of paleoclimates. Since the model includes a plant growth model that considers environmental factors such as water, O_2 and salt stress, it can simulate water consumption and recharge on a daily basis. Paleoclimate simulations could include temporal changes in rainfall distribution temperature, and subsequently plant water uptake.

Realistic representations of chemical weathering require not only knowledge of existing mineralogy and surface area but temperature, water inputs, evapotranspiration, hydraulic properties, and biological production and its influence on dissolution-enhancing ligands and soil CO_2 concentrations. To date, consideration of biological factors is minimal and probably the aspect most lacking in the existing models. In addition, predictive models will need to incorporate reaction rates rather than equilibria assumptions for most processes. This development is presently limited by the lack of kinetic information.

REFERENCES

Allison, J.D., D.S. Brown, and K.J. Novo-Gradac. 1990. MINTEQA2/PRODEFA2, a geochemical assessment model for environmental systems: Version 3.0. Office Res. Dev., USEPA, Athens, GA.

Amrhein, C., and D.L. Suarez. 1988. The use of a surface complexation model to describe the kinetics of ligand-promoted dissolution of anorthite. Geochim. Cosmochim. Acta 52:2785–2793.

Amrhein, C., and D.L. Suarez. 1992. Some factors affecting the dissolution kinetics of anorthite at 25 C. Geochim. Cosmochim. Acta 56:1815–1826.

Bales, R.C., and J.J. Morgan. 1986. Dissolution kinetics of chrysotile at pH 7 to 10. Geochim. Cosmochim. Acta 49:2281–2288.

Ball, J.W., E.A. Jenne, and M.W. Cantrell. 1981. WATEQ3: A geochemical model with uranium added. U.S. Geol. Surv. Open File Rep. 81-1183. U.S. Geol. Surv., Menlo Park, CA.

Ball, J.W., D.K. Nordstrom, and D.W. Zachmann. 1987. WATEQ4F—A personal computer fortran translation of the geochemical model WATEQ2 with revised data base. U.S. Geol. Surv. Open File Rep. 87-50. U.S. Geol. Surv., Menlo Park, CA.

Bennett, P.C., M.E. Melcer, D.I. Siegel, and J.P. Hassett. 1988. The dissolution of quartz in dilute aqueous solutions of organic acids at 25 C. Geochim. Cosmochim. Acta 52:1521–1530.

Blesa, M.A., A.J.G. Maroto, and A.E. Regazzoni. 1984. Boric acid adsorption on magnetite and zirconium dioxide. J. Colloid Interface Sci. 99:32–40.

Bowden, J.W., S. Nagarajah, N.J. Barrow, A.M. Posner, and J.P. Quirk. 1980. Describing the adsorption of phosphate, citrate and selenite on a variable charge mineral surface. Aust. J. Soil Res. 18:49–60.

Brown, D.S., and J.A. Allison. 1987. MINTEQA1, an equilibrium metal speciation model. EPA-600/3-87-012. Office Res. Dev., USEPA, Athens, GA.

Busenberg, E., and L.N. Plummer. 1982. The kinetics of dissolution of dolomite in CO_2-H_2O systems at 1.5 to 65 C and 0 to 1 atm. P_{CO_2}. Am. J. Sci. 282:45–78.

Casey, W.H., H.R. Westrich, G.W. Arnold. 1988. Surface chemistry of labradorite feldspar reacted with aqueous solutions at pH=2, 3, and 12. Geochim. Cosmochim. Acta 52:2795–2807.

Cederberg, G.A., R.L. Street, and J.O. Leckie. 1985. A groundwater mass transport and equilibrium chemistry model for multicomponent systems. Water Resour. Res. 21:1095–1104.

Chou, L., R.M. Garrels, and R. Wollast. 1989. Comparative study of the kinetics and mechanisms of dissolution of carbonate minerals. Chem. Geol. 78:269–282.

Davis, J.A., R.O. James, and J.O. Leckie. 1978. Surface ionization and complexation at the oxide/water interface. I. Computation of electrical double layer properties in simple electrolytes. J. Colloid Interface Sci. 63:480–499.

Davis, J.A., and J.O. Leckie. 1978. Surface ionization and complexation at the oxide/water interface. II. Surface properties of amorphous iron oxyhydroxide and adsorption of metal ions. J. Colloid Interface Sci. 67:90–107.

Davis, J.A., and J.O. Leckie. 1980. Surface ionization and complexation at the oxide/water interface. 3. Adsorption of anions. J. Colloid Interface Sci. 74:32–43.

Dudley, L.M., and R.J. Hanks. 1991. Model SOWACH: Soil-plant-atmosphere-salinity management model. Utah Agric. Exp. Stn. Res. Rep. 140.

Dzombak, D.A., and F.M.M. Morel. 1990. Surface complexation modeling. Hydrous ferric oxide. John Wiley & Sons, New York.

Felmy, A.R., D.C. Girvin, and E.A. Jenne. 1984. MINTEQ: A computer program for calculating aqueous geochemical equilibria. EPA-600/3-84-032. Office Res. Dev., USEPA, Athens, GA.

Furrer, G., and W. Stumm. 1986. The coordination chemistry of weathering: I. Dissolution kinetics of δ-Al$_2$O$_3$ and BeO. Geochim. Cosmochim. Acta 50:1847–1860.

Garrels, R.M., and C.L. Christ. 1965. Solutions, minerals and equilibria. Harper and Row, New York.

Goldberg, S. 1992. Use of surface complexation models in soil chemical systems. Adv. Agron. 47:233–329.

Goldberg, S., and G. Sposito. 1984. A chemical model of phosphate adsorption by soils. I. Reference oxide minerals. Soil Sci. Soc. Am. J. 48:772–778.

Harter, R.D., and G. Smith. 1981. Langmuir equation and alternate methods of studying "adsorption" reactions in soils. p. 167–182. In R.H. Dorody et al. Chemistry in the soil environment, ASA Spec. Publ. 40. ASA, Madison, WI.

Hayes, K.F., and J.O. Leckie. 1986. Mechanism of lead ion adsorption at the goethite-water interface. ACS Symp. Ser. 323:114–141.

Helgeson, H.C., W.M. Murphy, and P. Aagaard. 1984. Thermodynamic and kinetic constraints on reaction rates among minerals and aqueous solutions. II. Rate constants, effective surface area, and the hydrolysis of feldspar. Geochim. Cosmochim. Acta 48:2405–2432.

Hingston, F.J., and M. Raupach. 1967. The reaction between monosilicic acid and aluminum hydroxide. I. Kinetics of adsorption of silicic acid by aluminum hydroxide. Aust. J. Soil Res. 5:295–309.

Hohl, H., and W. Stumm. 1976. Interaction of Pb^{2+} with hydrous γ-Al$_2$O$_3$. J. Colloid Interface Sci. 55:281–288.

Hohl, H., L. Sigg, and W. Stumm. 1980. Characterization of surface chemical properties of oxides in natural waters. ACS Adv. Chem. Ser. 189:1–31.

Hoosbeek, M.R., and R.B. Bryant. 1992. Towards the quantitative modeling of pedogenesis—A review. Geoderma 55:182–210.

Hostetler, C.J., and R.L. Erikson. 1989. FASTCHEM package. Vol. 5. Rep. EA—5870-CCM. Elec. Power Res. Inst., Palo Alto, CA.

Inskeep, W.P., and P.R. Bloom. 1985. An evaluation of rate equations for calcite precipitation kinetics at pCO$_2$ less than 0.01 atm. and pH greater than 8. Geochim. Cosmochim. Acta 49:2165–2180.

Inskeep, W.P., and P.R. Bloom. 1986. Kinetics of calcite precipitation in the presence of water-soluble organic ligands. Soil Sci. Soc. Am. J. 50:1167–1172.

Jennings, A.A., D.J. Kirkner, and T.L. Theis. 1982. Multicomponent equilibrium chemistry in groundwater quality models. Water Resour. Res. 18:1089–1096.

Jury, W.A., H. Frenkel, and L.H. Stolzy. 1978. Transient changes in the soil-water system from irrigation with saline water. I. Theory, Soil Sci. Soc. Am. J. 42:579–585.

Kharaka, Y.K., W.D. Gunter, P.K. Aggarwal, E.H. Perkins, and J.D. DeBraal. 1988. SOLMNEQ.88: A computer program for geochemical modeling of water-rock interactions. U.S. Geol. Surv. Water Resources Invest. Rep. 88-4227. U.S. Gov. Print. Office, Washington, DC.

Kharaka, Y.K., and I. Barnes. 1973. SOLMNEQ: Solution-mineral equilibrium computations. U.S. Geol. Surv. Computer Contributions. Natl. Tech. Inform. Serv. no. PB-215899. U.S. Gov. Print. Office, Washington, DC.

Liu, C.W., and T.N. Narasimhan. 1989. Redox-controlled multiple-species reactive chemical transport. I. Model development. Water Resour. Res. 25:869–882.

Mangold, D.C., and C.F. Tsang. 1991. A summary of subsurface hydrological and hydrochemical models. Rev. Geophys. 29:51–79.

Marion, G.M., W.H. Schlesinger, and P.J. Fonteyn. 1985. CALDEP: A regional model for soil (Caliche) deposition in southwestern deserts. Soil Sci. 139:468–481.

Marion, G.M., and W.H. Schlesinger. 1994. Quantitative modeling of soil-forming processes in deserts: The CALDEP and CALGYP models. p. 129–145. In R.B. Bryant and R.W. Arnold (ed.) Quantitative modeling of soil forming processes. SSSA Specl. Publ. 39. SSSA, Madison, WI.

McDuff, R.E., and F.M. Morel. 1973. Description and use of the chemical equilibrium program RED-EQL2. Tech. Rep. EQ-73-02. California Inst. Technol., Pasadena, CA.

Nancollas, G.H., and M.M. Reddy. 1971. The crystallization of calcium carbonate. II. Calcite growth mechanism. J. Colloid Interface Sci. 37:824–829.

Papelis, C., K.F. Hayes, and J.O. Leckie. 1988. HYDRAQL: A program for the computation of chemical equilibrium composition of aqueous batch systems including surface-complexation modeling of ion adsorption at the oxide/solution interface. Tech. Rep. no. 306. Dep. Civil Eng., Stanford Univ., Stanford, CA.

Parkhurst, D.L., D.C. Thorstenson, and L.N. Plummer. 1980. PHREEQE. A computer program for geochemical calculations. U.S. Geol. Surv. Water Resour. Invest. Rep. 80-96. U.S. Gov. Print. Office, Washington, DC.

Plummer, L.N., T.M.L. Wigley, and D.L. Parkhurst. 1978. The kinetics of calcite dissolution in CO_2-water systems at 5–60 C. and 0.0 to 1.0 atm CO_2. Am. J. Sci. 278:179–216.

Plummer, L.N., E.C. Prestemon, and D.L. Parkhurst. 1991. An interactive code (NETPATH) for modeling net geochemical reactions along a flow path. U.S. Geol. Surv. Water Resour. Invest. Rep. 91-4-78. U.S. Gov. Print. Office, Washington, DC.

Robbins, C.W., R.J. Wagenet, and J.J. Jurinak. 1980. A combined salt transport chemical equilibrium model for calcareous and gypsiferous soils. Soil Sci. Soc. Am. J. 44:1191–1194.

Schaffer, M.J., and W.E. Larson. 1987. NTRM, a soil-crop simulation model for nitrogen, tillage, and crop residue management. USDA-ARS Conserv. Res. Rep. 34-1. Natl. Tech. Inform. Serv., Springfield, VA.

Schindler, P.W., B. Fürst, R. Dick, and P.U. Wolf. 1976. Ligand properties of surface silanol groups. I. Surface complex formation with Fe^{3+}, Cu^{2+}, Cd^{2+}, and Pb^{2+}. J. Colloid Interface Sci. 55:469–475.

Schindler, P.W., and H. Gamsjäger. 1972. Acid-base reactions of the TiO_2 (anatase)-water interface and the point of zero charge of TiO_2 suspensions. Kolloid-Z. Z. Polymere 250:759–763.

Schott, J. 1990. Modeling of the dissolution of strained and unstrained multiple oxides: The surface speciation approach. p. 337–365. In W. Stumm (ed.) Aquatic chemical kinetics. John Wiley & Sons, New York.

Schulz, H.D., and E.J. Reardon. 1983. A combined mixing cell/analytical model to describe two-dimensional reactive solute transport for undirectional groundwater flow. Water Resour. Res. 19:493–502.

Simunek, J., and D.L. Suarez. 1993. Modeling of carbon dioxide transport and production in soil. 1. Model development. Water Resour. Res. 29:487–497.

Sposito, G. 1984. The surface chemistry of soils. Oxford Univ. Press, Oxford, England.

Sposito, G. 1989. The chemistry of soils. Oxford Univ. Press. Oxford, England.

Sposito, G., and J. Coves. 1988. SOILCHEM: A computer program for the calculation of chemical speciation in soils. Kearney Found. Soil Sci. Univ. California, Riverside.

Sposito, G., and S.V. Mattigod. 1977. GEOCHEM: A computer program for the calculation of chemical equilibria in soil solutions and other natural water systems. Dep. of Soil & Environ. Sci., Univ. of California, Riverside.

Stumm, W., H. Hohl, and F. Dalang. 1976. Interaction of metal ions with hydrous oxide surfaces. Croat. Chem. Acta 48:491–504.

Stumm, W., R. Kummert, and L. Sigg. 1980. A ligand exchange model for the adsorption of inorganic and organic ligands at hydrous oxide interfaces. Croat. Chem. Acta 53:291–312.

Stumm, W., B. Wehrli, and E. Wieland. 1987. Surface complexation and its impact on geochemical kinetics. Croat. Chem. Acta 60:429–456.

Stumm, W., and R. Wollast. 1990. Coordination chemistry of weathering: Kinetics of the surface-controlled dissolution of oxide minerals. Rev. Geophys. 28:53–69.

Suarez, D.L. 1983. Calcite supersaturation and precipitation kinetics in the lower Colorado River, All-American Canal and East Highline Canal. Water Resour. Res. 19:652–661.

Suarez, D.L., and J. Simunek. 1992. Carbonate chemistry model with calcite kinetics combined with one-dimensional unsaturated water flow. p. 246. In Agronomy abstracts. ASA, Madison, WI.

Suarez, D.L., and J.D. Wood. 1991. Weathering of soil silicates isoluted from an arid zone soil. p. 252–253. In Agronomy abstracts. ASA, Madison, WI.

Suarez, D.L., J.D. Wood, and I. Ibrahim. 1992. Reevaluation of calcite supersaturation in soils. Soil Sci. Soc. Am. J. 56:1776–1784.

Svensson, U., and W. Dreybrodt. 1992. Dissolution kinetics of natural calcite minerals in CO_2-water systems approaching calcite equilibrium. Chem. Geol. 100:129–145.

Thrailkill, J. 1970. Solution geochemistry of the water of limestone terrains. Univ. Kentucky Water Resour. Inst. Res. Rep. 19, Univ. Kentucky, Lexington.

Truesdell, A.H., and B.F. Jones. 1974. WATEQ, A computer program for calculating chemical equilibria of natural waters. J. Res. U.S. Geol. Surv. 2:233–248.

van Riemsdijk, W.H., G.H. Bolt, L.K. Koopal, and J. Blaakmeer. 1986. Electrolyte adsorption on heterogeneous surfaces: Adsorption models. J. Colloid Interface Sci. 109:219–228.

Velbel, M.A., 1986. The mathematical basis for determining rates of geochemical and geomorphic processes in small forested watersheds by mass balance: Examples and implications. In S.M. Colman and D.P. Dethier (ed.) Rates of chemical weathering of rocks and minerals. Acad. Press, New York.

Wagenet, R.J., and J.L. Hutson. 1987. LEACHM—Leaching estimation and chemistry model. Center Environ. Res., Cornell Univ., Ithaca, NY.

Wagenet, R.J., J. Bouma, and J.L. Hutson. 1994. Modeling water and chemical fluxes as driving forces of pedogenesis. p. 17–35. *In* R.B. Bryant and R.W. Arnold (ed.) Quantitative modeling of soil forming processes. SSSA Spec. Publ. 39. SSSA, Madison, WI.

Westall, J.C. 1979. MICROQL. I. A chemical equilibrium program in BASIC. II. Computation of adsorption equilibria in BASIC. Tech. Rep. Swiss Fed. Inst. Technol., EAWAG, Dübendorf, Switzerland.

Westall, J. 1980. Chemical equilibrium including adsorption on charged surfaces. ACS Adv. Chem. Ser. 189:33–44.

Westall, J.C. 1982. FITEQL: A computer program for determination of equilibrium constants from experimental data. Rep. 82-01. Dep. Chem., Oregon State Univ., Corvallis, OR.

Westall, J.C. 1986. Reactions at the oxide-solution interface: Chemical and electrostatic models. ACS Symp. Ser. 323:54–78.

Westall, J.C., J.L. Zachary, and F.M.M. Morel. 1976. MINEQL: A computer program for the calculation of chemical equilibrium composition of aqueous systems. Tech. Note 18. Ralph M. Parsons Lab. Dep. Civil Eng., Massachusetts Inst. Technol., Cambridge, MA.

Yates, D.E., S. Levine, and T.W. Healy. 1974. Site-binding model of the electrical double layer at the oxide/water interface. J. Chem. Soc. Faraday Trans. I 70:1807–1818.

Yeh, G.T., and V.S. Tripathi. 1990. HYDROGEOCHEM: A coupled model of HYDROlogic transport and GEOCHEMical equilibria in reaction multicomponent systems. Rep. ORNL-6371. Oak Ridge Natl. Lab., Oak Ridge, TN.

Zinder, B., G. Furrer, and W. Stumm. 1986. The coordination chemistry of weathering: II. Dissolution of Fe(III) oxides. Geochim. Cosmochim. Acta 50:1861–1869.

4   Impact of Spatial Variabili֢ ֍ on Interpretive Modeling

L.P. Wilding

Texas A&M University
College Station, Texas

J. Bouma

Agricultural University
Wageningen, the Netherlands

Don W. Goss

USDA-SCS
Burleson, Texas

Pedology is the earth science that quantifies the factors and processes of soil formation including the quality, extent, distribution and spatial variability of soils from microscopic to megascopic scales (Sposito & Reginato, 1992). Spatial variability is governed by the processes of soil formation which are in turn interactively conditioned by lithology, climate, biology, and relief through geologic time. Pedologists have long recognized spatial variability as the mainstay of their profession but are being strongly challenged to better integrate this information for assessments of land quality, risk assessment, and environmental protection and developing nonagriculture interpretations. Pedologists have developed qualitative landscape models heavily premised on the state-factor analysis approach (Jenny, 1941, p. 281). Applicability and limitations of this conceptual model have been recently critiqued (Wilding, 1994). Soils are welded together into a continuum like chains—processes and impacts on higher topographic surfaces directly affect adjacent lower lying surfaces. This is because transfer of energy flow and mass flux, the driving forces of pedogenesis, occur within and over three-dimensional soil landform bodies.

Pedology is an integrative and extrapolative science. It provides an organizational framework to catalogue modes, mechanisms and magnitudes of spatial variability, and to generalize this knowledge base for synthesis of models. It thus provides a vehicle for extrapolation and scaling of spatial variability from components of soils (hand specimens and horizons) to the population of soils within the continuum as a whole (pedons, toposequences, physiographic entities and the pedosphere) (Fig. 4–1).

Spatial variability is the change in a soil property as a function of time and space. It may be temporal or of a more permanent nature, but it is a real landscape

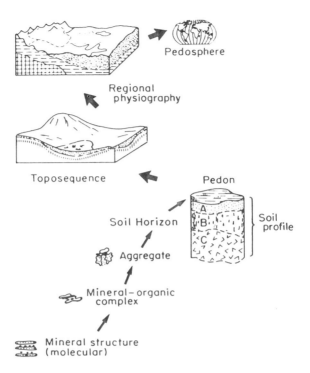

Fig. 4–1. Soil components and systems as various hierarchical levels (from Sposito & Reginato, 1992).

attribute—the norm. The challenge for soil scientists is to convey this body of knowledge to modelers in a format that transmits magnitude, pattern and form (Nordt et al., 1991). The soil variability dilemma is that soils are a continuum, many properties are not single valued, many properties are temporal, and properties are systematically and time spatially dependent. In short the medium is vertically and laterally anisotropic.

Numerous symposia and a voluminous body of literature have addressed soil and landscape spatial variability attributes and methods to quantify this characteristic (e.g., Beckett & Webster, 1971; Webster, 1977; Miller, 1978; Arnold, 1979; Wang, 1982, p. 1–34; Wilding & Drees, 1983; Nielsen & Bouma, 1985; Upchurch et al., 1988; Mausbach & Wilding, 1991). There continues to be strong inertia to develop more comprehensive models of soil biometrics driven largely by the era of information and technology explosion, availability of computer hardware and software, and a systems analytical approach. Further, simulation models of plant growth, water and chemical solute transport, land quality, climatic change and environmental quality are heavily based on soil properties and pedogenic processes (Finke, 1992, p. 131; Kirchner, 1992; Vanclooster et al.,

1992). Hence, Bryant and Olson (1987) and Hoosbeek and Bryant (1992) have urged pedologists to place greater emphasis on development of simulation models for future pedogenic and landscape modeling efforts. Waltman et al. (1992) have recently published the *Proceedings of the First Soil Genesis Modeling Conference* as a step in this direction.

The purposes of this paper are to summarize modes, magnitudes and forms of spatial variability in soil systems and to explore consequent impacts on modeling. Emphasis in this paper will be interpretive use of pedologic data in terms of modeling crop growth, water regimes and the associated land qualities.

SOIL VARIABILITY

Scale and Origin of Spatial Variability

Spatial variability in soil systems belongs to two broad categories—systematic (structured) and random (unstructured and unknown causes). Systematic variability is a gradual or marked change in soil properties as a function of physiography, geomorphology and interactions of soil-forming factors (Wilding & Drees, 1983). Systematic variability occurs at submicroscopic to megascopic scales. Systematic variation permits pedologists to partition spatial variability in soils by subsets of properties that constitute soil survey map units corresponding to geomorphic landscape elements (summit, shoulder, backslope, footslope, etc.). Close-interval (large-scale) spatial variability of a systematic nature may be as great or greater than long-range interval changes (small scale). An example of this are shrink-swell phenomena in soils that give rise to gilgai topographic relief variability in physical, and corresponding subsoil chemical and biological, properties at intervals of meters or less (Wilding et al., 1991).

At yet more refined scales (hand specimens, aggregate ped units and microfabrics), systematic organization of organic and inorganic constituents occurs as coatings of clay along void surfaces, zonation of oxyhydroxides, and concentrations of soluble salts and carbonates within the soil matrix and along conductive voids (Wilding & Hallmark, 1984). Even microorganisms are systematically organized about the rhyzosphere of roots, about clay colloids and along ped interfaces (Sposito & Reginato, 1992). These distribution patterns reflect hydraulic flow, diffusion, immobilization and microbial colonization processes at micron and submicron scales in soil systems.

Causes of vertical and lateral anisotropy that yield spatial variability of a random nature over short-range or indeterminate distances include: differential lithology, differential hydrology; differential intensity of pedogenic weathering processes; differential biological activity and pedoturbation; differential erosion and sedimentary accretion; temporal effects of soil management; and sampling and analytical errors. All of the above, except the latter two, may contribute to systematic variation, but the effects may be too subtle or complex to be discerned visibly or by measurement.

The purpose of soil surveys is to partition the spatial variability of landforms into stratified subsets that are less variable than the medium as a whole (Fig. 4–2)—it is to remove systematic components of error. It is important to note,

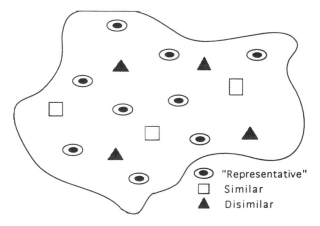

Fig. 4–2.   Schematic illustration of the composition of a map unit delineation portraying observation of "representative," similar and dissimilar soils within the mapped area.

however, that appreciable variability still remains in mapping units of soil series (cartographic units) used to partition real geomorphic landscape components (Wilding & Drees, 1983; Mausbach & Wilding, 1991). Considerable data have been amassed by field soil scientists and the National Cooperative Soil Survey (NCSS) program that demonstrate that cartographic units of landscapes rarely comprise more than 40 to 50%, and sometimes less than 20% of the soils named in the map unit (Mokma, 1987; Nordt et al., 1991). This is not so serious as it may first appear, however, because the interpretative purity for most of these units is 50 to 85%, at the 80% probability level (Nordt et al., 1991). This means that while spatial variability of properties within a given delineated map unit area may be substantial, as long as it is within the limits that control similar use and management behavior, the absolute variability of soils may be of less concern. Current NCSS standards for map unit composition require at least 75% of the soils comprising the map unit to have similar interpretative ratings (Soil Survey Staff, 1983).

Magnitude of Soil Property Variability

Users of soil surveys and modelers frequently wish to know the relative magnitude of soil property variability. Table 4–1 illustrates means and ranges in coefficients of variability (CV) that have been reported in the literature for a select number of soil properties sampled from equivalent horizons or depths within landscape mapping units of the same soil series (Upchurch et al., 1988). While these are only guidelines, they serve as useful indices in the absence of on-site data (Table 4–1).

In considering these properties, their relative magnitudes of variability, and operational procedures used in conducting soil surveys, modelers are reminded of the following expected trends:

1.   Reliability in estimating soil properties decreases with depth because relatively few observations are made at depths >2 m in standard soil survey operations—limited ground truth.

Table 4–1. Relative variability of selected soil properties sampled within mapping units of a given soil series.

Soil property	CV(%)† Mean	CV(%)† Range	Relative order of soil variability
Bulk density	7	5–13	
Soil color hue	9	2–20	
Soil color value	10	4–12	Least variable
Soil pH	10	5–15	
Plasticity limit	15	5–28	
Liquid limit	17	8–31	
A Horizon thickness	18	8–31	
Water retention (33 kPa)		10–31	
Base saturation	25	17–33	
Total sand content	25	8–46	Moderately
Total clay content	25	10–61	variable
Calcium carbonate equivalence	28	20–30	
Soil color chroma	28	15–50	
Depth to carbonates	30	20–49	
Cation exchange capacity	32	20–40	
Depth to mottling	35	20–50	
Organic matter content	39	20–61	
Plasticity index	41	20–63	
Soil thickness	43	25–58	Most
Exchangeable Ca	48	30–73	variable
Exchangeable K	57	7–160	
Exchangeable Mg	58	31–121	
Water-soluble salt extract	48	–	
Hydraulic conductivity	75	13–150	

†The coefficient of variability (CV) values represent variations for equivalent horizons or depths.

2. More permanent (stable) soil properties such as soil texture, mineralogy soil thickness, and color are less variable than temporal or more dynamic properties such as water content, hydraulic conductivity, redox state, salt content, biological activity, exchangeable cations and organic matter content (Fig. 4–3).

3. Spatial variability in soils increases with the nature of the parent materials in the following order: loess < till < fluvial deposits < pyroclastic and tectonic rocks < drastically disturbed materials.

4. Properties which are measured and closely calibrated to a standard (e.g., texture, color, pH, etc.) are less variable than qualitatively accessed parameters (soil structure, consistency, porosity, root abundance, etc.).

At the sampling unit level, spatial variability over an area of 1 to 2 m is generally reduced relative to that expressed in a landform for the same properties. The CV's for more stable properties range from about 5 to 10%, while for the more dynamic ones, they commonly range from 10 to 20%, with extremes up to 35% (Wilding & Drees, 1983). Laboratory error analysis is property dependent but commonly with CV's less than 5% (Wilding & Drees, 1983). The relative variability as a function of size of the units sampled is schematically illustrated in Fig. 4–4.

66 WILDING ET AL.

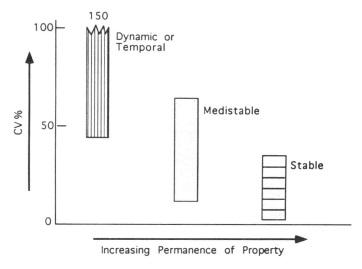

Fig. 4–3. The relative variability of soil properties (attributes) as a function of the permanence of the property.

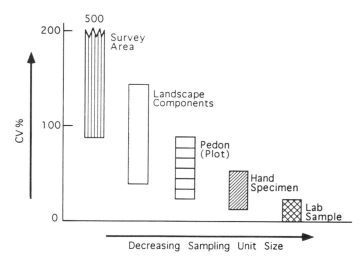

Fig. 4–4. The relative variability of soil properties (attributes) as a function of size of the sampling unit.

Current Efforts to Accommodate Variability into Models

Approaches to accommodate spatial variability into models range widely among disciplines and by nature of the model. Engineers accommodate modeling uncertainty by incorporating an arbitrary safety factor into the results of the simulation model. Many modelers do not attempt to incorporate error analysis, or will argue the system is too complex (Hoffman, 1991). Others argue that the purpose of a process-based model is not to predict results but rather to identify voids or shortcomings in the model construct; hence no error analysis is deemed necessary or appropriate. Still others contend that a model is not valid until calibrated with sensitivity and error analyses. In the following section several approaches are described by soil scientists to accommodate spatial variability into models.

Delineated Areas on the Soil Map as Expressions of Spatial Variability

Several modeling attempts have been reported using soil map delineations as a means to express spatial variability patterns, using "representative" soil profiles for each delineation. Bouma et al. (1980a,b) calculated moisture supply capacities with a capacity-type simulation model for average weather conditions during the growing season for Dutch conditions. Such calculations have become standard soil survey interpretation procedures in the Netherlands for calculating the effects of lowered water table levels on yield of agricultural crops. Comparable developments occurred in Denmark (Hansen et al., 1991) and Belgium (Vereecken et al., 1990). The implicit assumption that delineated areas on a soil map can serve to adequately represent spatial variability patterns in an area of land is, of course, not correct because considerable variation may occur within the delineated areas. However, before this aspect will be analyzed, it is necessary to further specify the character of spatial variability. When interpreted in terms of common soil characteristics (such as clay and organic matter content color, bulk density, etc.) variability among different soil units may be considerable (see previous discussion and Table 4–1). This, however, does not necessarily imply an identical degree in variability in the type of soil characteristics or qualities that a user will be interested in. More specifically, Wösten et al. (1985) reported variability of calculated moisture supply capacities for soil series that were considerably lower than variability among the primary soil data of the different soil series, represented on the soil map. The soil map had 350 delineations—the derived map showing areas with different classes of moisture supply capacity had only 100 units. In other words, spatial variability in terms of pedological features may be significantly different from the spatial variability in terms of some functional feature, such as the moisture supply capacity, which is of interest to a user.

Recently a symposium, "Soil Assemblies for Water/Chemical Transport-Productivity and Management," met in Atlanta, Georgia, to discuss soil assemblies for models. The symposium, discussed systems for developing soil assemblies to reduce the number of determinations that must be made to evaluate the impact of soil-chemical-climate interactions on the environment. One soil in a class of the assemblies would represent the impact of all the soils in that class. Several unique methods of accomplishing this goal were presented, of which some have not been published.

The Fertility Capability Classification System (FCC) (Sanchez et al., 1982), as modified by Dyke and Fauchs (Texas A&M Blackland Exp. Stn., Temple, TX), was used to classify each soil series in the USA using the modified FCC. The series were grouped by FCC and a series was selected which would be representative of a group of series with similar productivity and management. The selected soils were used in a simulation model to provide soil production and erosion output for national policy analysis.

Quinsenberry et al. (1993) have developed an initial version of a soil classification system for water and chemical transport through soils. The current system was developed for South Carolina and groups soils into eight classes. The classification system identifies soil properties that affect the water and chemical transport process, and will provide the framework grouping soils for input into models that incorporate these same properties. Larry West (Univ. of Georgia) and Clint Truman (ARS, Tifton, GA) expanding the work by Leonard et al. (1989) used 50 soils from Georgia for GLEAMS simulations (Leonard et al., 1987) of leaching and runoff of 10 different pesticides. Average amounts of pesticide loss were used to group soils that behave similarly. The properties and classification of soils in the groups were evaluated to define groups of soils for GLEAMS input that would have similar output.

Spatial Variability within Mapping Units

The functional analysis, which implies interpretations of "representative profiles" for a wide range of applications, still focuses implicitly on mapping units as homogeneous "carriers" of information.

The question as to how variability within mapping units can be characterized has been discussed by various authors (Nordt et al., 1991; Burrough, 1991). Stein et al. (1991) raised the basic issue as to whether soil data should be "averaged" first, followed by calculations (the common procedure, discussed above) or whether calculations should be made for point data first, followed by interpolation to areas of land. They showed that best predictions were obtained with the second procedure. In another paper, Stein et al. (1988) showed that such interpolations were most accurate when made within major delineated areas of the soil map, rather than in arbitrarily delineated areas, such as gird configurations. Thus, available soil maps are useful to improve interpolation. By making model calculations for point data and by using interpolation techniques, such as different forms of kriging, predictions can be obtained for areas of land including an estimate of the reliability involved. Thus a statistically defined expression is obtained for spatial variability, which has proven very valuable for environmental soil studies dealing with soil contamination (e.g., Staritsky et al., 1992). Of particular interest here is disjunctive kriging which expresses results in terms of probabilities of exceeding certain limiting values (e.g., Yates et al., 1986; Finke & Stein, 1993). In areas without clear spatial structures, use of geostatistics is irrelevant because only nugget effects can be distinguished. Then classical statistics may well be used providing estimates of averages and their standard deviations, which, again, express spatial variability.

In using soil survey data, special note should be made of when the field work was conducted, when the survey was published, scale of mapping, changing soil concepts with time, field methodology used, accuracy of estimates reported, etc. The soil survey should not be used beyond the limits of its constraints. It is not intended to provide site-specific information. A field soil scientist should be consulted if several different survey areas are involved for continuity and accuracy of the data base.

How to Express Field Experience on Variability in Models

Simulation modeling is being applied ever more widely as models become more user-friendly and computers become more powerful and less expensive. Besides, model output appears to have an invincible aura to many users. There are, however, serious dangers in applying models indiscriminatory. Flow of water is usually represented by Darcian flow principles which may be unrealistic in many soils with large pores or stratified horizons. The interaction between physical, chemical and biological processes is infinitely more complex in nature than can be simulated by even the most complex simulation model. It is therefore crucial to collect field data for calibration and validation of models. Modern noninvasive remote sensing techniques (e.g., ground penetrating radar, electromagnetic induction, infrared videography, etc.) can be used as well as in situ monitoring equipment. But, in addition, ways should be found to express field experience by trained soil survey personnel. Using soil map delineations as a means of stratification for applying geostatistical interpolation techniques is one specific example of using expert knowledge before applying modeling (Stein et al., 1988). But a more fundamental approach to the application of simulation modeling may be needed.

Bouma (1993) proposed a thorough analysis of questions asked by users of soil information, followed by an application of expert knowledge to determine whether the question raised could be answered by existing information. In this view, simulation modeling would only be applied for those questions where current knowledge appears to be inadequate. This approach was advocated in reaction to what seems to be unquestioned application of models to any perceived or real problem. In many cases, such applications do not represent efficient use of available resources including the wide body of expert knowledge in many areas in soil science. A specific example of this mixed qualitative/quantitative approach was presented by Van Lanen et al. (1992) when studying potato (*Solanum tuberosum* L.) growth in the Netherlands. The expert system ALES (Van Wambeke et al., 1992) was used to screen out less suitable sites. Simulation was focused on sites with a relatively high potential.

Expression of Parameter Variability in Model Calculations

Use of "representative" soil profiles to characterize mapping units, represents a case where soil properties are "averaged" first, followed by calculations for that particular profile which is assumed to be representative for the particular soil series and associated mapping units of the soil map. As discussed previously, an alternative is to make calculations for each point observation and apply interpolation to obtain expressions for land areas. The only errors involved here are the experimental errors related to the estimation or measurement of model parameters. For example, calculation of the soil moisture supplying capacity requires knowledge of moisture retention and hydraulic conductivity data. They can be measured, but measurements are usually rather expensive and time consuming. An alternative is to use estimates based on soil characteristics that can be obtained easily from soil survey data. Estimates can be derived with regression analysis (e.g., Cosby et al., 1984). Functions relating available soil data to parameters needed for simulation, have been called pedotransfer functions (Bouma, 1989). They can be continuous pedotransfer functions, such as the regression equations of Cosby et al. (1984) or class pedotransfer functions which relate physical or chemical parameters to particular soil horizons (Wösten et al., 1985; Finke et al., 1991). Continuous pedotransfer functions represent covarying property relationships derived from large data bases such as the National Soil Survey Information System (NASIS) in the USA where algorithms can be developed between measured and predicted property values. It is reasonable to expect that data become less accurate when estimated, rather than when directly measured. Class pedofunctions would represent category groupings of the population into entities such as horizons and layers, diagnostic horizons, different kinds of parent material, soil series, mapping units or phases of soil series, and other systems of property cataloguing such as the fertility capability classification system. Wösten et al. (1990) compared results of simulations of the water regime of soils using different types of continuous and class pedotransfer functions. In their study, differences obtained, as compared with independently measured values, were not significantly different. The example illustrates that variability of basic hydraulic characteristics, as such, is less critical than the variability obtained when these characteristics are used in a model to calculate soil qualities that are of interest to the user. The same conclusion was drawn earlier when comparing variability of basic soil data and land qualities.

Simulations for point data using measured values have a variability that is determined by the experimental error involved in the measurement and associated calculation procedures. Different experimental methods have different errors which is insufficiently realized when choosing methods, because manuals only tend to emphasize technical and theoretical aspects of methods, rather than operational ones (e.g., Bouma, 1983). Be that as it may, the variability of hydraulic characteristics within a given area or among points sampled within one particular soil series, can be expressed by not only using an average curve for, as an example, the moisture retention or hydraulic conductivity, but a series of curves which represent the observed variability. Model calculations can use a series of random choices from the curves presented by, for example, Monte-Carlo type methods, to generate output. Petach et al. (1991) used this approach to express variability of pesticide leaching for soils in the northeastern USA. Bouma and Hackten Broeke (1993) applied this procedure by using extreme values of a series of experimental curves to obtain variabilities of soil moisture supply capacity, trafficability and aeration status for an identical soil under grassland and different tillage management practices When calculations are made for a number of years, an expression can be obtained for variability in space and time.

The two examples mentioned above varied only some parameters for the simulation model. Ideally, all input parameters should be expressed in terms of their variability to allow generation of model input that would truly represent spatial and temporal variability. A massive computing capacity would be needed to allow such procedures. It is therefore advisable to start with a sensitivity analysis to find out which parameters are particularly sensitive and to focus the variability analysis on these parameters. In fact, such a sensitivity analysis applied to the models themselves might indicate dominant processes that overwhelm less sensitive processes, thereby reducing model complexity and commensurate data demands.

AREAS OF FUTURE NEEDS

Soil Survey Data Bases for Modeling

It is generally agreed by most modelers and soil scientists that error analysis of simulation outputs is critical to model calibration and validation. It also is probable that the accuracy of future simulations will be constrained by accurate input data (Wösten et al., 1990). Use of available data requires that we are aware of the uncertainty of our predictions. This is especially critical with the voluminous National Cooperative Soil Survey (NCSS) data base presented in soil survey reports when one considers the many estimated rather than measured properties. The NASIS is developing a strategy to enhance and improve the collection, storage, manipulation and dissemination of soil survey information in a manner to facilitate decision making. It will do this by: (i) support field activities to efficiently gather data bases, (ii) application of expert knowledge to make information usable for an increasing variety of purposes, and (iii) making information readily available for a wider clientele of users. It is important to the success of NASIS and to the NCSS program to have strategies reviewed by modelers to be certain the data base will serve simulation modeling purposes. The NASIS provides the information to document spatial variability of landscape entities. It serves as the data base to drive "soil assemblies for models" previously discussed. In this context, it is critical for the assessment of spatial variability of landforms and the development of pedogenic process models that the spatial diversity elements for mapping units not be averaged but rather represented as individual observations for transects and soil composition analysis of landscape units. Comprehensive data sets to quantify basic soil properties (e.g., clay content, pH, CEC, hydraulic conductivity, organic matter content, etc.) within landscape units is labor intensive and cost restrictive. Hence, in the near term, landscape variability will be estimated or inferred from soil survey information systems. With few exceptions, comprehensively documented research sites, as reported by Bouma and colleagues in the Netherlands, cannot be justified. However, their pioneering work establishes a basis for greater utilization of soil survey tabular and spatial information than previously contemplated. They have demonstrated how spatial variability, which is commonly viewed in a negative posture, may actually enhance the ability of models to achieve realistic field-scale scenario analysis (Finke, 1992, p. 131).

Comprehensive use of the soil survey data bases for modeling will rely heavily on the development of continuous and class pedotransfer functions. The class-pedotransfer functions are particularly relevant for soil surveys, because soil horizons are well defined in a landscape context. Future efforts will therefore need to focus on assembling the data base so that given horizons or layers commonly utilized by pedologists to catalogue vertical and lateral anisotropy in soils is assembled by groupings that convey process behavior or quality attributes. Perhaps this assembly may be by diagnostic horizons or subgroupings of same for given taxa. The challenge will then be to identify a few key morphogenetic properties that covary with pedogenic processes so simulation models developed on comprehensive "hard data sets" may be interpolated and the simulations extended to sites for which "hard data sets" are not available. Statistical tools such as multiple regression, multivariant analyses and cokriging will enhance this effort. The ultimate goal will be to couple morphogenetic properties that reflect pedogenic processes with simulation models that reflect the driving forces for pedogenesis—water movement and mass flux. Hydraulic behavior simulation models will be the core for such simulations because they link water movement, solute transfer, suspended translocation of colloids, organic matter production, and oxidation-reduction environments. Diffusion models also will be needed to reflect dynamics of transformation and weathering processes. Disjunctive kriging and cokriging may be powerful tools to examine the extreme events that drive pedogenesis such as probability of exceeding water flux of a given magnitude to a given soil depth or associated transfer of colloids and solutes.

Geostatistics and Model Parameter Sensitivity

The objective of modeling landscapes is to estimate some particular impact on the landscape, surface or groundwater. Spatial variability of model input parameters within the landscape being modeled is important to understand. However, of equal or greater importance is the sensitivity of model input parameters to model output. A parameter with a large variance across the landscape may have little sensitivity to the model output. A soil parameter that has a small variance across the landscape may have a high sensitivity to model output, and thus produce a high variability in model output. These two factors, landscape variability and model sensitivity, weigh strongly in determining what parameters should be measured.

Single Values for Soil Parameters

Most models currently in use require soil parameters not commonly measured or single values. Many such attributes are currently represented in soil data bases as ranges. Universally accepted pedotransfer functions are not available to estimate the unmeasured parameters. The National Cooperative Soil Survey Program in the USA will shortly release a soils data base and computer program that generates numerous pedotransfer functions that will be important to the soil modeler. The generated values include moisture volume-tension data, saturated hydraulic conductivity, and porosity-moisture data. The data base will also contain single values for other soil parameters represented as ranges in the Soil Conservation Society soils interpretations record data base.

Communicating the Variability Problem

Study of soil spatial and temporal variability is seen by many as a rather academic analysis and not a crucial element of soil science. Pedology has traditionally dealt with spatial variability, but expressions of factors of soil formation and their spatial variability in quantitative terms are largely perceived as being of academic interest. Modern uses of soil information, however, put the variability question in a quite different perspective. Environmental regulations and laws increasingly specify critical concentrations of potential pollutants in both soil and groundwater. Whether these contents are exceeded or not at a given location, may be a question with major economic repercussions. The very character of soil variability makes it absurd to define, for example, a critical nitrate content in groundwater of 50 mg/L, because large variations in contents occur in the field at short distances. As discussed, new interpolation techniques, such as disjunctive kriging, allow predictions in terms of probabilities of exceedance rather than as a particular content with a defined accuracy (Yates et al., 1986). This development is intriguing and represents, in fact, a realistic and relevant expression of variability in practical terms as was recently demonstrated when characterizing the spatial effects on groundwater quality of a series of fertilizing scenarios, using simulation techniques (Finke, 1992, p. 131).

SUMMARY

In summary, soil scientists know more about landscapes and soil property (attribute) variability than has been conveyed to modelers, or incorporated into simulation models. Spatial variability that is commonly viewed in a negative posture may actually enhance the ability to assess the error analysis in models and develop field-scale scenarios of pedogenesis and soil behavior. The challenge is to get scientists with soil-based expertise and those with modeling expertise together so field and modeling experiences may be coupled.

This combination of field truth and modeling will enhance the identification of a modeling approach that provides a quantitative estimate of system behavior from a simple and relatively inexpensive data set while also providing an indication of the uncertainty of predictions. This approach suggests that the model should provide input data that integrate basic soil properties as soil assemblies or pedotransfer functions [representative series, profile(s) or horizon(s) that carry common behavioral attributes] rather than to utilize the soil properties per se as input data. Regression analysis can be used to develop algorithms between the basic soil properties and functional elements of system behavior. The rationale for this approach is that the interaction between physical, chemical and biological processes in soils is infinitely more complex in nature than can be simulated by even the most complex simulation model. In such approaches, it is imperative to collect field data for calibration, verification and error analysis of the simulated results for at least case examples of the population being modeled.

Further, for modeling of pedogenic processes, regional hydrology expressing recharge, discharge and through-flow processes should be coupled with hydraulic simulation models at given points in the landscape (Richardson et al., 1992). Challenges of developing pedogenic simulation models is an ambitious undertaking but one that is likely to reap rich rewards in the future. It is worth the risk of failure.

REFERENCES

Arnold, R.W. 1979. Strategies for field resource inventories. Agron. Mimeo 70–120. Dep. Agron., Cornell Univ., Ithaca, New York.

Beckett, P.H.T., and R. Webster. 1971. Soil variability: (A review). Soils Fert. 34:1–15.

Bouma, J. 1983. Use of soil survey data to select measurement techniques for hydraulic conductivity. Agric. Water Manage. 6:177–190.

Bouma, J. 1989. Using soil survey data for quantitative land evaluation. p. 225–239. In B.A. Stewart (ed.) Advances in soil science. Vol. 9. Springer Verlag, New York.

Bouma, J. 1993. Soil behavior under field conditions: Differences in perception and their effects on research. Geoderma 60:1–14.

Bouma, J., and M.J.D. Hackten Broeke. 1993. Simulation modeling as a method to study land qualities and crop productivity related to soil structure differences. Geoderma 57:51–67.

Bouma, J., P.J.M. de Laat, R.H.C.M. Awater, H.C. van Hessen, A.F. van Holst and Th. J. van de Nesl. 1980a. Use of soil survey data in a model for simulating regional soil moisture regimes. Soil Sci. Soc. Am. J. 44:808–814.

Bouma, J., P.J.M. de Laat, A.F. van Holst, and Th.J. van de Nesl. 1980b. Predicting the effects of changing water-table levels and associated soil moisture regimes for soil survey interpretations. Soil Sci. Soc. Am. J. 44:797–802.

Bryant, R.B., and C.G. Olson. 1987. Soil genesis: Opportunities and new directions for research. p. 301–311. In L.L. Boersma et al. (ed.) Future developments in soil science research. SSSA, Madison, WI.

Burrough, P.A. 1991. Sampling designs for quantifying map unit composition. p. 89–127. In M.J. Mausbach and L.P. Wilding (ed.) Spatial variabilities of soils and landforms. SSSA Spec. Publ. 28. SSSA, Madison, WI.

Cosby, B.J., G.M. Hornberger, R.B. Clapp, and T.R. Ginn. 1984. A statistical exploration of the relationships of soil moisture characteristics to the physical properties of soils. Water Resour. Res. 20:682–690.

Finke, P.A. 1992. Spatial variability of soil structure and its impact on transport processes and some associated land qualities. Ph.D. diss. Wageningen Agricultural Univ., Wageningen, Netherlands. p. 131.

Finke, P.A., J. Bouma, and A. Stein. 1991. Measuring field variability of disturbed soils for simulation purposes with geostastical and morphological techniques. Soil Sci. Soc. Am. J. 56:187–192.

Finke, P.A., and A. Stein. 1993. Application of disjunctive co-kriging to optimize fertilizer additions on a field scale. Geoderma (In Press).

Hansen, S., H.E. Jensen, N.E. Nielsen, and H. Svendsen. 1991. Simulation of nitrogen dynamics and biomass production in winter wheat using the Danish simulation model DAISY. Fert. Res. 27:245–259.

Hoffman, O. 1991. Conclusions of BIOMOVS phase I. Proc. Meet. Validity Environ. Transfer Models, Swedish Radiation Protection Inst., Stockholm, Sweden.

Hoosbeek, M.R., and R.B. Bryant. 1992. Towards the quantitative modelling of pedogenesis—A review. Geoderma 55:183–210.

Jenny, H. 1941. Factors of soil formation. McGraw-Hill Book Co., Inc., New York.

Kirchner, G. 1992. A method of modeling the transfer of radio nuclides deposited after the Chernobyl accident via the grass-cow-milk pathway. Model. Geo-Biosphere Proc. 1:13–21.

Leonard, R.A., W.G. Knisel, and D.A. Still. 1987. GLEAMS: Groundwater loading effects of agricultural management systems. ASAE 30:1403–1418.

Leonard, R.A., H.F. Perkins, and W.G. Knisel. 1989. Relating agrichemical runoff and leaching to soil taxonomy: A GLEAMS model analysis. p. 158–160. In K.J. Hatcher (ed.) Proc. 1989 Georgia Water Resour. Conf. Univ. Georgia, Athens, GA. 16–17 May.

Mausbach, M.J., and L.P. Wilding. 1991. Spatial variabilities of soils and landforms. SSSA Spec. Publ. 28. SSSA, Madison, WI.

Miller, F.P. 1978. Soil surveys under pressure: The Maryland experience. J. Soil Water Conserv. 33:104–111.

Mokma, D.L. 1987. Soil variability of five landforms in Michigan. Soil Surv. Land Eval. 7:25–31.

Nielsen, D.R., and J. Bouma. (ed.) 1985. p. 2443. In Soil spatial variability. Proc. Workshop ISSS, SSSA, Las Vegas, NV. 30 Nov.–1 Dec. 1984. Pudoc, Wageningen, the Netherlands.

Nordt, L.C., J.S. Jacob, and L.P. Wilding. 1991. Quantifying map unit composition for quality control in soil survey. p. 183–197. In M.J. Mausbach and L.P. Wilding (ed.) Spatial variabilities of soils and landforms. SSSA Spec. Publ. 28. SSSA, Madison, WI.

Petach, M.C., R.J. Wagenet, and S.D. de Gloria. 1991. Regional water flow and pesticide leaching using simulations with spatially distributed data. Geoderma 48:245–269.

Quinsenberry, V.L., B.R. Smith, R.E. Phillips, H.D. Scott, and S. Nortcliff. 1993. A soil classification system for describing water and chemical transport. Soil Sci. 156:306–315.

Richardson, J., L.P. Wilding, and R.B. Daniels. 1992. Recharge and discharge of groundwater in the aquic moisture regime illustrated with flow net analysis. Geoderma 53:65–78.

Sanchez, P.A., W. Couto, and S. Boul. 1982. The fertility capability soil classification system: interpretation, applicability, and modification. Geoderma 27:283–309.

Soil Survey Staff. 1983. National soils handbook. U.S. Gov. Print. Office, Washington, DC.

Sposito, G., and R. Reginato. 1992. Opportunities in basic soil science research. SSSA, Madison, WI.

Staritsky, I.G., P.H.M. Sloot, and A. Stein. 1992. Spatial variability and sampling of cyanide polluted soil on former galvanic factory premises. Water Air Soil Pollut. 61:1–16.

Stein, A., M. Hoogerwerf, and J. Bouma. 1988. Use of soil map delineations to improve (co)kriging of point data on moisture deficits. Geoderma 43:163–177.

Stein, A., I.G. Staritsky, J. Bouma, A.C. Van Eynsbergen, and A.K. Bregt. 1991. Simulation of moisture deficits and areal interpolation by universal co-kriging. Water Resour. Res. 27:1963–1973.

Upchurch, D.R., L.P. Wilding, and J.L. Hatfield. 1988. Methods to evaluate spatial variability. p. 201–229. In L.R. Hossner (ed.) Reclamation of disturbed lands. CRC press, Boca Raton, FL.

Vanclooster, M., H. Vereecken, J. Diels, F.H. Huysmans, W. Verstraete, and J. Feyen. 1992. Effect of mobile and immobile water in predicting nitrogen leaching from cropped soils. Model. Geo-Biosphere Proc. 1:23–40.

Van Lanen, H.A.J., M.J.D. Hackten Brocke, J. Bouma, and W.J.M. de Groot. 1992. A mixed qualitative/quantitative physical land evaluation methodology. Geoderma 55:37–54.

Van Wambeke. A.R., and D.G. Rossiter. 1992. Automated land evaluation system (ALES). Vers. 3. SCAS Teach. Ser. no. 2. Dep. Soil, Crop & Atmospheric Sci., Cornell Univ., Ithaca, New York.

Vereecken, H., M. van Clooster, and N. Swerts. 1990. A model for the estimation of nitrogen leaching with regional applicability. p. 250–263. In R. Merckx et al. (ed.) Fertilization and the environment. Leuven Univ. Press, Leuven, Belgium.

Waltman, W.J., E.R. Levine, and J.M. Kimble. 1992. Proc. 1st soil genesis modeling conf. USDA-SCS, Nat. Soil Surv. Center, Lincoln, NE. 13–15 Aug. 1991.

Wang, C. 1982. Application of transect method to soil survey problem. Land Resour. Res. Inst. Contrib. 82-02. LRRI, Res. Branch, Agric. Canada, Ottawa, Ontario.

Webster, R. 1977. Quantitative and numerical methods in soil classification and survey. Clarendon, Oxford.

Wilding, L.P. 1994. Factors of soil formation: Contributions to pedology. p. 15–30. In R. Amundson et al. (ed.) Factors of soil formation: A fiftieth anniversary retrospective. SSSA Spec. Publ. 33. SSSA, Madison, WI.

Wilding, L.P., and L.R. Drees, 1983. Spatial variability and pedology. p. 83–116. In L.P. Wilding et al. (ed.) Pedogenesis and soil taxonomy: I. Concepts and interactions. Elsevier Publ. Co., Amsterdam.

Wilding, L.P., and C.T. Hallmark. 1984. Development of structural and microfabric properties in shrinking and swelling clays. p. 1–22. In J. Bouma and P.A.C. Raats (ed.) Proc. ISSS Symp. Water Solute Movement in Heavy Clay Soils, Wageningen, the Netherlands. ILRI Publ. 37. Wageningen, the Netherlands.

Wilding, L.P., D. Williams, W. Miller, T. Cook, and H. Eswaran. 1991. In Characterization, classification, and utilization of cold aridisols and vertisols. p. 232–247. Proc. of 6th Int. Soil Correlation Meet. USDA-SCS, Natl. Soil Surv. Center, Lincoln, NE.

Wösten, J.H.M., J. Bouma, and G.H. Stoffelsen. 1985. The use of soil survey data for regional soil water simulation models. Soil Sci. Soc. Am. J. 49:1238–1245.

Wösten, J.H.M., C.H.J.E. Schuren, J. Bouma, and A. Stein. 1990. Comparing four methods to generate soil hydraulic functions in terms of their effect on simulated soil water budgets. Soil Sci. Soc. Am. J. 54:832–837.

Yates, S.R., A.W. Warrick, and D.E. Myers. 1986. Disjunctive kriging. 1 Overview of estimation and conditional probability. Water Resour. Res. 22:615–621.

5　A Comprehensive Framework for Modeling Soil Genesis

E.R. Levine and R.G. Knox

NASA/Goddard Space Flight Center
Greenbelt, Maryland

Simulation modeling provides a means by which the general knowledge and assumptions that occur in a particular scientific subfield can be collected, integrated, and tested. Among its many advantages are: (i) defining the state-of-the-art understanding of a specific system, (ii) piecing together components of complex systems in a unified manner to create a theoretical construct of the total system, (iii) exercising and testing theories to determine how well they hold true quantitatively within the paradigm of present knowledge, (iv) identification of strengths and weaknesses in scientific knowledge which can be used to direct future research, and (v) making predictions of the state of the system over space and time.

Long-standing limitations to simulation modeling on digital computers include approximating continuous processes with discrete mathematics, the difficulty of verifying whether complex programs behave as specified, and a mismatch between the conceptual objects in scientific theory and the data flows and structures of traditional software design (cf., Fishman, 1978). Numerical programming and modern software engineering have made significant advances in all these areas. Particularly significant for modeling complex heterogeneous systems has been a shift in paradigms in software engineering, from processors and data flows to objects and behavior (see Winblad et al., 1990). Object-oriented software can more easily mimic the way science decomposes and views a complex system.

The need to model the soil environment and soil genesis processes has become more critical as the scientific community assesses effects of global change on various ecosystems. Soils play a key role in controlling the structure and function of ecosystems and will act both as a buffer to global change as well as be affected by changes. In order to understand the soil's contribution from an ecological standpoint, the specific processes occurring within the soil itself as well as the relationship between the soil and other components of the ecosystem must be examined. Theoretical constructs of soils and theory relationships to their surrounding environment have always been used in soil science and pedology, and these historical paradigms must be used as the basis on which to build more quantitative models.

THE HISTORY OF SOIL MODELING

Historically, models of soil genesis tended to be qualitative and descriptive. From an ecosystem point of view, the soil system can be characterized as the "pedosphere" which exists at a junction and develops as a result of interaction between the lithosphere, atmosphere, hydrosphere and biosphere of the planet (Arnold et al., 1990). Figure 5–1 shows the role of the pedosphere as an integrator of the interaction and functioning of the other geospheres. From this perspective, the pedosphere regulates the exchange of matter and energy between geospheres, allowing certain substances and fluxes of energy to pass through, while reflecting, retaining, or accumulating others at its surface, or within the soil profile.

The best-known qualitative model of pedologic processes was described by Jenny (1941) as "the factors of soil formation." In this model, the interaction between topography (landforms), parent material (underlying geology), climate (temperature and moisture), and biologic activity (flora and fauna including humans) occurring through time (soil age) define how the system properties evolve from their initial state (Jenny, 1941; Jenny, 1980). This model is used extensively at every level of soil science from field survey for soil mapping to creating digital layers used in geographic information systems. Because of its simple yet comprehensive use of basic state factors that govern soil development, this model provides the fundamental framework for quantifying soil processes. Using field and laboratory measurements of soil properties controlled by the state factors, the next step in the evolution of soil modeling is to produce a

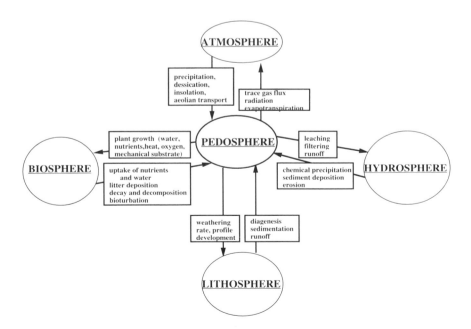

Fig. 5–1.   The role of the pedosphere in the global climate system.

mathematical description of the soil system showing how the mechanisms operating within the soil system interact.

Hoosbeek and Bryant (1992) give a good overview of many of the important qualitative and quantitative approaches to soil genesis modeling that have been developed. Most quantitative soil models developed to date focus on discrete chemical, physical, or mineralogical processes and isolate these mechanisms from other processes occurring simultaneously in the soil. Many of the most quantitative models focus solely on a single system (e.g., the leaching of chemicals through the soil profile, geochemical transformations of chemical species, or the physics of heat and water transport through the soil). These are effective in that they highlight individual soil processes and describe them by mechanistic, stochastic, or empirical algorithms so that they can provide simple predictions of system states. However, they generally do not provide a description of the soil from a pedogenic perspective since certain mechanisms must be generalized or omitted for the purpose of simplifying the model. Even Jenny, in attempting to solve the state factor equations, showed that simple equations did not yield significant results, and that soils had to be stratified and integrated in a way that would take into account the complex interactions between soil properties in order to develop meaningful algorithms (Jenny, 1980).

QUANTITATIVE MODELING OF PEDOLOGIC PROCESSES

Modeling soil forming processes quantitatively becomes more difficult than modeling more restricted sets of processes because a number of requirements must be met to adequately describe the total soil system. A soil genesis simulation model must be comprehensive in order to capture the interactions of the pedosphere with the other geospheres, interpret the exchange between each of the five soil forming factors in producing soil profile characteristics, and describe the complex mechanisms which operate in a "process response" manner in which a change in the property in one part affects a process or property in another (Yaalon, 1975). The great spatial variability of soils both on a landscape level, and with depth, as well as the wide range of temporal scales at which soil processes are operating add additional complicating factors that must be considered. A comprehensive model would have to consist of parameters and conditions that were dynamic in order to reflect changes over a variety of time scales. In addition, a model of soil genesis should make use of the great wealth of existing soil characterization data as sources of input data as well as for verification and testing of model simulation results.

While these requirements may seem rigorous and perhaps unrealistic, recent scientific and technological advances have made quantitative modeling of soil forming processes more feasible. These advances and their role in quantitatively modeling soil processes follow.

The Development of Soil Taxonomy

Soil Taxonomy (Soil Survey Staff, 1975) has provided a method for emphasizing common properties of soils which allow them to be grouped hierarchically, and thus, characterize their variability at a given scale. In this way, chemical, physical and morphological information about a given soil profile becomes available depending on the degree of detail limited by which level of the taxonomic scheme the users information occurs. For example, at the series level, detailed quantitative information from laboratory analysis can be obtained about a particular soil that can be used as input for simulation models. At higher taxonomic categories such as subgroups, great groups, or up through orders, more general information is available that can, at least, provide ranges of quantitative data for model input (e.g., base saturation, depth to lithic or paralithic contact, amount of organic matter, etc.). While this information is more general, it is superior to omitting soil properties completely from models because of lack of data, or making incorrect sweeping inferences about soil properties that are not based on some realistic source of soils information. By categorizing soils with Soil Taxonomy, even the most general information from sources such as soil maps can be used to derive quantitative data about a given soil.

Soil Data Base Compilation

The collection and organization of comprehensive soil data bases providing quantitative data on physical, chemical, mineralogical, and morphological characteristics has greatly enhanced resources for developing quantitative relationships describing soil properties and processes. A large quantity of excellent field and laboratory data have been collected by the USDA Soil Conservation Service (SCS) as well as various other U.S. and international organizations and universities (e.g., Cunningham et al., 1972; FAO-UNESCO, 1978: USDA-SCS, 1987; Van Waveren & Bos, 1988; Kirkwood et al., 1989). These data provide information that can be readily accessed for use in modeling exercises. Their standardized field and laboratory methods also serve as a prototype for obtaining comparable information at other sites. Within a soil model itself, algorithms must be developed to make use of these data as input for driving mechanistic simulations of the soil system. Examples of these type of algorithms are given in the model by Levine and Ciolkosz (1986, 1988). Driving variables include climate parameters such as precipitation and air temperature, whereas boundary conditions and state variables included thickness, particle size, bulk density, percentage rock fragments, cation exchange capacity, exchangeable cations (Ca, Mg, Na, and K), percentage organic C and N, percentage Fe and aluminum oxide, soil pH, clay mineralogy, and percentage water holding capacity at one-third (0.03 MPa) and 15 bar (1.5 MPa) tensions for each horizon.

As these properties change over time from their starting values within the model, new values must be recalculated using mechanistic, stochastic, or empirical algorithms to simulate soil processes. Some properties that are boundary conditions in short-term simulations (e.g., thickness, percentage rock fragments) become evolving state variables in very long-term pedologic simulations.

In some cases, the types of data available from standard soil characterization may be incomplete, or information about dynamically important soil properties may not be routinely measured in the field or laboratory. Under these circumstances, surrogate values can be predicted from empirical relationships, artificial intelligence techniques, or process algorithms derived from available measurements. For example, soil C stocks under forests are poorly approximated by C measurements made on soil profiles sampled under field crop conditions. In this situation, C estimates under forested soils may be determined through extrapolation from the limited amount of data that may be available to adequately describe C status over a large region. Another example is the use of theoretical relationships between parameters such as aluminum oxide and sulfate adsorption, or particle size and available water holding capacity to predict these variables when data for one or the other is not available (Petersen et al., 1968; Barrow, 1970). These types of estimates can later be used to build mechanistically based algorithms that can better describe and predict the property for which little data presently exists.

Spatial Data Products and Geographic Information System

In conjunction with digitized soil characterization data bases, digitized maps at various scales also are presently available. These maps show the spatial distribution of soil types as well as specific properties that correspond with attributes obtained from characterization data that can be displayed as thematic maps. With the availability of digital soils information, a means for organizing and manipulating this data is critical. The development of Geographic Information System (GIS) techniques allows spatial and temporal analysis of digitized information. With GIS, layers of information are combined so that they can be georeferenced for display, statistical analysis, or used as a source of data for specific soil attributes at a given location to use as input for simulation modeling.

The State Soil Geographic Data Base (STATSGO) (1:250:000) and National Soil Geographic Data Base (NATSGO) (1:1:000:000) produced by the USDA Soil Conservation Service (USDA-SCS, 1991) as well as the Canadian Land Potential Data Base (1:1:000:000) (Kirkwood et al., 1989) are examples of this type of digitized map product. Both the STATSGO and NATSGO maps as well as the Canadian map have been used to calculate and display estimates of the soil organic C pool at their respective scales (Bliss et al., 1994; Tarnocai et al., 1993). Overlaying these C estimates with other data layers such as climate, vegetation type, topography, or remotely sensed imagery can provide a means for both spatial and temporal modeling leading to the development of algorithms to explain C status in soils under various conditions (Levine et al., 1994; Merry & Levine, 1994).

Technological Advances in Computing Machinery

Advances in the design and performance of digital computers continue to expand what is feasible using numerical simulation. Personal computers of today have capabilities seen only in high-end technical work stations a short time ago, and today's technical work stations can tackle difficult, large memory problems that were formerly restricted to supercomputers. For a decade, the numerical performance of workstation computers has been doubling every 12 to 18 mo (cf., Gwennap, 1992a,b). The practical limits to this trend have not yet appeared (Gwennap, 1992b). In the interim, disciplines such as statistics have made numerical simulation a routine tool of most researchers, rather than the preserve of a computationally literate minority (Kempthorne et al., 1991). Among the implications of this phenomenon are that system simulations of "supercomputer" levels of complexity can be undertaken by a decentralized community without routine access to premium large-scale computers.

"Object-Oriented" Modeling Techniques

Advances in computer software also provide grounds for optimism about quantitative, process-based simulation of soil genesis. Modern operating systems available on computers of moderate cost can support distributing computations across numerous processors without a great deal of specialized programming. Fundamentally, the same revolution in software engineering that provided the tools to design and build some of these complex operating systems and user interfaces also can be applied to scientific simulation. Design strategies pioneered for building user interfaces in the Smalltalk language and in the area of simulation to support programming in the Simula language (Wegner, 1990) have been formalized as object-oriented software engineering (e.g., Meyer, 1988; Booch, 1991).

In an approach called "object modeling technique" (Rumbaugh et al., 1991) "objects" corresponding to physical entities or concepts are grouped together into classes of similar structure, and classes define manipulations and information that will be passed from one module to another. A range of required object behavior is then coded into a coherent software "object" rather than in routines that perform similar operations on a variety of different types of data. Objects having the same structure are grouped together in a "class." The class definition then specifies the data structure and "member functions" or "methods" permitted to manipulate the data objects of that type. Rather than passing around data structures, modules in object-oriented programs typically pass "messages" to each other specifying what manipulations to perform. The design of an object then specifies what it does when receiving a particular sort of message. For example, an object-oriented data base could return a requested data value, without the requesting routine needing any information about the underlying data structures. Although the terminology and notation varies among authors, the underlying concepts are similar (Jacobson, 1993).

The benefits of object-oriented software development include a large measure of independence from specific programming languages, data encapsulation, reduced sensitivity to program modification, rapid test and evaluation prototyping, and the reuse of basic modeling modules. This approach focuses on overall system design for optimal comprehensibility and ease of modification, rather than data flows or optimization for a particular computer. Leading advocates of object-oriented software design (e.g., Booch, 1991) emphasize comprehensibility and ease of modification as goals to reduce the overall costs of writing and maintaining complex software, but they also are vital goals for scientific simulation if scientists who are not computer professionals are to participate in developing and testing parts of complex simulation programs.

Object-oriented programming languages (e.g., Smalltalk, CLOS, C++, Objective-C) provide special support to make some kinds of object-oriented designs easier to implement. However, it is possible and often advantageous to implement an object-oriented design in a procedural or functional language (Rumbaugh et al., 1991). Older languages such as C, Ada, or Fortran 77 may provide greater robustness or portability (Rumbaugh et al., 1991; Booch, 1991). For example, a complex simulation system might mix C or C++ for the user interface and system calls, with Fortran for numerical portions.

A GENERAL FRAMEWORK FOR SOIL GENESIS MODELING

By combining all of these capabilities, a quantitative approach can be taken to simulation modeling of soil genesis which will provide a description of the initial conditions in a soil profile, predict changes that occur over time under various environmental conditions, and determine effects of new soil properties on other components of the ecosystem.

Using the above tools, a general framework for quantitatively modeling soil processes can be built. Key characteristics of such a framework must include the following:

1. Flexibility in simulating time steps. A variety of time steps is required to adequately simulate the hierarchy of rates at which soil processes operate. Changes in soils can be divided into three time scales: the short-term scale (daily and seasonal), the mesoscale (10–100 yr), and the long-term (the pedologic time scale of thousands of years). Short-term properties reflect the present dynamic soil condition, and will be reflected in the immediate resources available to support physiological functioning of vegetation or land use. Mesoscale properties are those that are currently of interest in questions of global change in that transformations of these types determine the ability of the soil to act as a substrate for vegetative health or succession, yet can be managed or affected to produce a given property or affect. Long-term properties change over centuries through geologic time scales, and determine the genesis and morphology of a given soil profile, and thus, its long-term potential to support plant growth. Table 5–1 lists some of the important soil properties imbedded in the proposed soil genesis simulation and the time scale at which they can change.

    At the same time that properties within soil horizons change, the processes associated with these properties also vary. Transformations in mineralogy, for example, are pedologically slow mechanisms, and therefore, are not obvious unless long-term simulations are run. Short-term processes, such as changes in solution chemistry will vary within each minimum time step depending on the input of precipitation and corresponding exchange mechanisms within each soil horizon. Results produced by medium-time scale processes, such as those related to organic matter content (bulk density, nutrient and water holding capacity) are sensitive at a "meso-term" time step and length of simulation run. Thus, the soil submodel must be designed to accommodate inquiries about soil processes and properties that span very fine to long-term temporal scales in order to understand both the direct effect of the soil on the forest

Table 5–1. Rates of change of some soil properties.†

Short term (daily to seasonal)
Temperature
Moisture
Gas phase composition
Mesoterm (annual to decadal)
pH
Nutrient status
Organic matter composition
Microbiota
Bulk density
Porosity
Infiltration
Water holding capacity
Structure
Sesquioxide content
Long term (centuries to geologic time)
Mineral composition
Particle size distribution
Particle density

†Adapted from Richter (1987) and Arnold et al. (1990).

ecosystem as well as its long-term potential to provide a substrate for vegetative growth.

2. A comprehensive description of processes and interactions. Figure 5–2 is an example of a comprehensive soil model framework. Major soil processes are divided into five clusters: (i) abiotic chemical reactions, (ii) activities of organisms, (iii) energy balance and water phase transitions, (iv) hydrologic flows, and (v) particle redistribution (Fig. 5–2). Each of the columns listed within the first row of Fig. 5–2 represent groupings of major mechanisms operating during pedogenesis. The clusters listed represent distinct classes of soil processes that are often the main focus of different types of simulation models (e.g., soil physics, soil chemistry, or soil mineralogy models). Although not shown in this figure, each of the individual processes listed are intimately linked with those in other columns, so that the state of the system at any given time is defined by the attributes which have been modified by the actions of the process clusters.

The second row contains the primary processes operating within horizons. The third row represents transfers between soil horizons and soil units, and the fourth row is an approximate range in time step of the listed processes. The fifth row of Fig. 5–2 lists attributes which can be measured and could be predicted as results from the soil model.

As an example of the interrelatedness of the components of this model structure, one can trace a component such as soil temperature. Soil thermal fluxes are key drivers of changes within soils. Temperature also controls rates of reactions within the soil (e.g., organic decomposition and nutrient uptake). The temperature of the soil also is critical for determining its ability to provide a suitable substrate for plant growth.

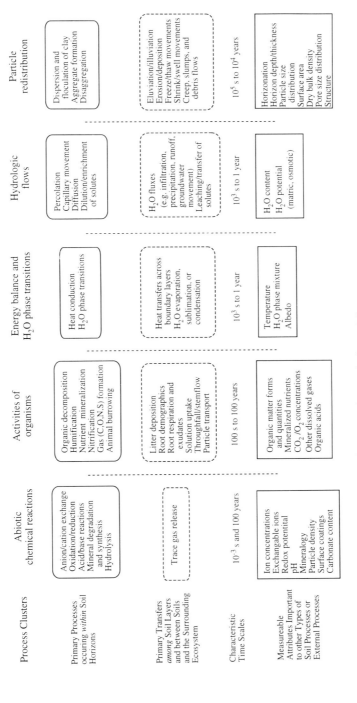

Fig. 5–2. Conceptual framework of important soil processes for quantitative modeling.

Soil properties are affected by temperature under varying conditions over time and at different time scales, while thermal fluxes occur at daily or less than daily rates. Using thermal model algorithms, predictions of soil temperature at various depths within the soil profile can be made based on soil characterization data for a particular profile. These data can then be made available when required in other process modules within the soil model, and in this way, a quantitative, dynamic representation of the soil system would be produced.

3. Modularity. In developing a simulation model of soil processes, the spatial and temporal resolution of each module must be independent from the others. Similarly, a user should be able to substitute different algorithms for specific processes without having to simultaneously revise the allocation and bookkeeping for all other processes that are affected by the results of that algorithm. As knowledge of soil processes increases, soil scientists also should be able to revise and improve submodels in their area of expertise, without those changes adversely affecting the behavior of other submodels. These scientists further should be able to perform model experiments, varying detail or spatial and temporal scales, without extensive technical assistance.

Using these requirements and the schematic in Fig. 5–2, a soil genesis simulation model could be built to operate with a modular structure. Within this structure, nine primary modules could be defined (see Fig. 5–3). These are: (i) five distinct modules simulating processes related to the major soil processes, (ii) a main program that acts as an event scheduler for processes and reporting over time, and (iii) three service modules for control and reporting which will handle interaction with scientists running the model and produce needed periodic and snapshot reports. Reporting intervals, simulation length, and temporal resolution for component processes would be under user control to the maximum extent feasible. Soil horizons would be modeled as sets of horizontal grid cells expressed in a common coordinate system. In this way, the processes need not be represented all at the same spatial scales.

SUMMARY AND CONCLUSIONS

Understanding the role of soils and soil processes in global change is critical as we move into the next century because of the integrative role soils play within all ecosystems, and the dependency of other "geospheres" on the pedosphere. Theoretical modeling of soils is not a new concept, and these qualitative paradigms of soil genesis and its relationship to the surrounding ecosystem must be used as the basis for quantifying specific processes. With these theoretical paradigms in mind, and the use of existing soil data, a comprehensive and quantitative approach can be taken in modeling soil genesis which can simulate the complexity of the interactions within the soil system.

Simulation modeling of pedogenesis is a viable method for combining many of these important interactions. Using software advances such as "object-oriented" programming, soil scientists can begin to conceive of the soil system as an integrated whole that can be described quantitatively. Specific processes can

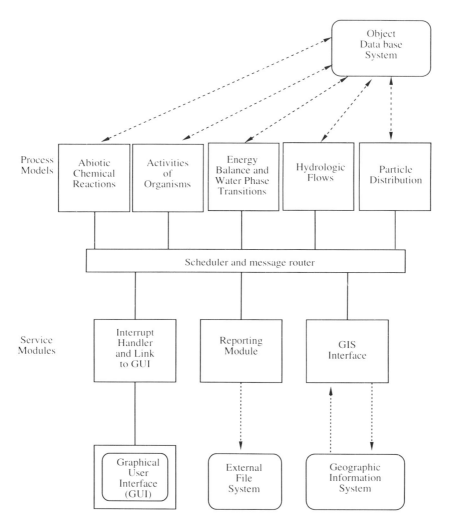

Fig. 5–3.   Design of a program implementing a soil process model showing control of execution for the primary program modules.

be viewed as "clusters" which operate individually but are still affected by and are critically linked to each of the other clusters in the system. Algorithms describing mechanisms defined by clusters can be developed from state-of-the-art research, or can drive new research initiatives where major gaps in understanding are identified. Using object-oriented programming, algorithms and routines can be easily updated and exchanged as new knowledge becomes available. Data is accessible to test models developed with this structure from a variety of sources including generalizations derived from Soil Taxonomy, specific characterization data, or digitized maps at many scales.

The final result of a simulation is a soil profile with properties that have changed from their initial conditions based on drivers derived from the state factors parent material, climate, topography, vegetation, and time. With a simulation tool that is flexible and comprehensive, scientists will be able to understand the soil system and predict how it will affect and be affected by many different scenarios of global change.

REFERENCES

Arnold, R.W., I. Szabolcs, and V.O. Targulian. 1990. Global soil change. IIASA, Laxenburg, Austria.

Barrow, N.J. 1970. Comparison of the adsorption of molybdate, sulfate, and phosphate by soils. Soil Sci. 109:282–288.

Bliss, N.B., S.W. Waltman, G.W. Petersen, and R.L. Day. 1994. Preparing a soil carbon inventory for the United States using geographic information systems. In R. Lal et al. (ed.) Soils and global change. Springer Verlag, New York. (In Press.)

Booch, G. 1991. Object-oriented-design with applications. Benjamin/Cummings Publ. Co., Redwood City, CA.

Cunningham, R.L., G.W. Petersen, R.P. Matelski, R.W. Ranney, and E.J. Ciolkosz. 1972. Laboratory characterization and field descriptions of selected Pennsylvania soils. Agron. Ser. 25. Pennsylvania State Univ., Univ. Park, PA.

Fishman, G.S. 1978. Principles of discrete event simulation. John Wiley & Sons, New York.

Food and Agriculture Organization—United Nations Education Scientific and Cultural Organization. 1975. Soil map of the world. FAO-UNESCO, Paris, France.

Gwennap, L. 1992a. Escape from the temple of doom. Microproc. Rep. 6:20–22.

Gwennap, L. 1992b. Microprocessor developments beyond 1995; Experts see limits to superscalar benefits—Memory becomes paramount. Microproc. Rep. 6:25–27.

Hoosbeek, M., and R. Bryant. 1992. Toward the quantitative modeling of pedogenesis—A review. p. 31–46. W.J. Waltman et al. (ed.) 1st Soil Genesis Modeling Conf., Lincoln, NE. 13–15 Aug. 1991. USDA-SCS, Natl. Soil Surv. Lab., Lincoln, NE.

Jacobson, I. 1993. Time for a cease-fire in the methods war. J. Object-Orient. Program. 6:6–84.

Jenny, H. 1941. Factors of soil formation. McGraw-Hill, New York.

Jenny, H. 1980. The soil resource, origin and behavior. Ecol. Stud. 37. Springer Verlag, New York.

Kempthorne, P., N. Mukhopadhyay, P.K. Sen, and S. Zacks. 1991. Research—How to do it: A panel discussion. Stat. Sci. 6:149–169.

Kirkwood, V., J. Dumanski, A. Bootsma, R.B. Stewart, and R. Muma. 1989. The land potential data base for Canada. User's handbook. Tech. Bull. no. 1983-4E. Agric. Canada, Ottawa, Ontario.

Levine, E.R., and E.J. Ciolkosz. 1986. A computer simulation model for soil genesis applications. Soil Sci. Soc. Am. J. 50:661–667.

Levine, E.R., and E.J. Ciolkosz. 1988. Determination of soil sensitivity to acid rain by computer simulation. Soil Sci. Soc. Am. J. 52:209–215.

Levine, E., R. Knox, and W. Lawrence. 1994. Relationships between soil properties and vegetation at the Northern Experimental Forest, Howland, Maine. Remote Sens. Environ. 47:231–241.

Merry, C., and E. Levine. 1994. Methods to assess soil carbon using Ground based techniques and remote sensing. R. Lal et al. (ed.) In Soils and global change. Springer Verlag, New York. (In Press.)

Meyer, B. 1988. Object-oriented software construction. Prentice-Hall, New York.

Petersen, G.W., R.L. Cunningham, and R.P. Matelski. 1968. Available moisture within selected Pennsylvania soil series. Agron. Ser. no. 3. Pennsylvania State Univ., Univ. Park, PA.

Richter, J. 1987. The soil as a reactor. Catena Verlag, Cremlingen, Germany.

Rumbaugh, J., M. Blaha, W. Premerlani, F. Eddy, and W. Lorensen. 1991. Object-oriented modeling and design. Prentice-Hall, Inc., New Jersey.

Soil Survey Staff. 1975. Soil taxonomy: A basic system of soil, classification for making and interpreting soil surveys. USDA-SCS. Agric. Handb. 436. U.S. Gov. Print. Office, Washington, DC.

Tarnocai, C., J.A. Shields, and B. MacDonald (ed.). 1993. Soil carbon data for Canadian soils (Interim Rep.) Contrib. 92–179. Centre Land and Biol. Resour. Res., Ottawa, Canada.

U.S. Department of Agriculture—Soil Conservation Service. 1987. National soil data base. Lincoln, NE.

U.S. Department of Agriculture—Soil Conservation Service. 1991. State soil geographic data base (STATSGO): Data users guide. USDA-SCS Misc. Publ. 1492. U.s. Gov. Print. Office, Washington, DC.

Van Waveren, E.J., and A.B. Bos. 1988. ISRIC soil information system, user manual. Tech. Pap. no. 15. Int. Soil Ref. Inform. Centre, Wageningen, the Netherlands.

Wegner, P. 1990. Concepts and paradigms of object-oriented software construction. Prentice-Hall, New York.

Winblad, A.L., S.D. Edwards, and D.R. King. 1990. Object-oriented software. Addison-Wesley Publ. Co., Reading, MA.

Yaalon, D.H. 1975. Conceptual models in pedogenesis: Can soil-forming functions be solved? Geoderma 14:189–205.

6

Measuring and Modeling Min Weathering in an Acid Forest Soil, Solling, Germany

L.G. Wesselink

Agricultural University
Wageningen, the Netherlands

G. Grosskurth

University of Göttingen
Göttingen, Germany

J.J.M. van Grinsven

National Institute of Public Health
 and Environmental Protection
Bilthoven, the Netherlands

Mineral weathering is an important nutrient source in terrestrial ecosystems, especially in ecosystems where pools of exchangeable base cation have been depleted, e.g., by soil acidification or intensive agricultural use. Weathering of parent material also strongly determines soil formation (Jenny, 1941) and groundwater quality. So, quantification of weathering rates at field conditions is of importance to: (i) assess the vulnerability of terrestrial ecosystems to human, disturbances and (ii) understand and predict the time scales of soil formation.

There are several approaches to quantify rates of mineral weathering. At the field scale one may estimate "historical" weathering from the bulk chemistry of the soil; mineral pools are compared with the composition of a reference C horizon or parent rock, taking a relatively stable mineral as an internal standard (Tarrah, 1988; Flehmig et al., 1990; Olsson & Melkerud, 1991). Weathering rates obtained with this method usually reflect thousands of years of weathering, but the environmental conditions during the past thousands of years may have been quite different from those today. Also, if a C horizon is used as reference, its weathering status is often unknown. Present field weathering rates can be estimated from element budgets at the scale of a soil profile or catchment. Weathering rates are calculated from the balance of atmospheric input, output by net uptake in biomass, and output with drainage or stream water. At best, these studies extend over two decades (Matzner, 1989; Kirchner, 1992), however, they face considerable uncertainties in the input flux of dry atmospheric deposition and in changes in the pools of exchangeable cations.

At the field scale, manipulation and control of the soil system is difficult, especially when a slow process like mineral weathering has to be studied. Therefore, the most widely applied method to assess weathering rates, is to decrease the spatial scale to a soil column, or to batch experiment in the laboratory. Often, the temporal scale of observation decreases concurrently. In the laboratory, experimental conditions can be controlled and a system can be manipulated to obtain information on rates and mechanisms of mineral weathering. A number of methods for studying weathering in the laboratory presented in the literature (for a review see van Grinsven & van Riemsdijk, 1992) have focused on the role of pH (Sverdrup, 1990), organic molecules (Manley & Evans, 1986), flow rate (van Grinsven & van Riemsdijk, 1992), mineral pretreatment (Holdren & Berner, 1979) and surface characteristics (Anbeek, 1992). However, rates obtained from laboratory experiments are generally one to three orders of magnitude higher than rates obtained from field studies (e.g., Schnoor, 1990).

Appreciating the advantages and disadvantages of the several approaches to assess weathering rates, we remain with the question of how to explain the high rates of mineral weathering under laboratory conditions, and their discrepancies with rates observed in the field. Such an explanation could lead to a method to extrapolate weathering rates obtained at the laboratory scale to the field situation.

In this paper, we determine rates and mechanisms of silicate weathering from an acid forest soil in the German Solling area (Ellenberg, 1971; Matzner, 1989). Forests ecosystems in this area are strongly impacted by acid deposition. As a result the soils at Solling have been depleted in exchangeable base cations (base saturation <4%), which emphasizes the importance of quantifying base cation weathering rates. The soil under investigation is located in an experimental forest where the chemistry of bulk precipitation, throughfall and soil solution has been measured since 1969 (Matzner, 1989). At Solling, field observations also include repeated measurements of pools of exchangeable cations. Using this unique long-term data set we derive field weathering rates by calculating input-output mass balances. Furthermore, 'historical' weathering rates are estimated from elemental analysis of the soil profile.

Next, the effect of soil depth, solution chemistry (e.g., pH), mineral saturation and temperature on mineral weathering, as determined in a number of laboratory experiments, is discussed. All these factors are considered to influence weathering rates in the field and are therefore potential parameters of a predictive weathering model. The weathering experiments focus on release of K and Mg from illite, the main reactive mineral source of these elements at Solling. The laboratory experiments are used to estimate parameters of proposed rate equations that describe release of Mg and K from illite. Subsequently, these rate equations are used to calculate K and Mg release at field conditions, and predictions are compared with the rates inferred from field methods.

THEORETICAL DESCRIPTION OF WEATHERING KINETICS

A general expression for the rate of mineral dissolution can be derived from transition state theory (Aagard & Helgeson, 1982; Nagy et al., 1991)

$$R_w = k \prod (a_i)^{m_i} \left(1 - e^{\left[\frac{n\Delta G}{RT} \right]^n} \right)$$ [1]

R_w is the dissolution rate, k is a rate constant, a_i is the activity of the Species i in solution, ΔG is the energy of the overall reaction, R the gas constant and T the temperature, m and n are constants. For dissolution of silicates below pH 5, the first term in Eq. [1] reduces to a_H^{mH} (Aagard & Helgeson, 1982). Eq. [1] can be rewritten to

$$R_w = k \left(H^+ \right)^m \left| 1 - \left(\frac{IAP}{Q_{eq}} \right)^n \right|$$ [2]

where Q_{eq} is the equilibrium constant and IAP the ion activity product for the dissolution reaction. Equation [2] was applied to illite weathering by van Grinsven et al. (1986), it describes a proton-dependent detachment rate of an activated complex on the mineral surface, with decreasing rate of detachment as solutions approach equilibrium with respect to a mineral phase (Nagy et al., 1991). Here Q_{eq} will be defined as the equilibrium product for the dissolution of a hypothetical Mg-Al-Si illitic phase

$$Mg_a Al_b Si_{3.5}(s) + (2a + 3b)H^+ \rightarrow aMg^{2+} + bAl^{3+} + 3.5Si$$

with

$$Q_{eq} = \frac{[Mg]^a [Al]^b [Si]^{3.5}}{[H]^{2a+3b}}$$ [2a]

Diffusion controlled release of K from illite interlayers can be described by Fick's first law of diffusion: $J = D\partial x/\partial c$. The cumulative K release (M_K) during a time interval is equal to the time integral of J. For steady state diffusion it is written

$$M_k = {}_0^t \int D \frac{\partial c}{\partial x} = k' \left(K_{crit} - K_{bulk} \right) t$$ [3a]

For nonsteady-state diffusion (Luce et al., 1972)

$$M_k = {}_0^t \int D \frac{\partial c}{\partial x} = k'' \left(K_{crit} - K_{bulk} \right) t^{0.5}$$ [3b]

In both equations, a critical potassium level (K_{crit}) is defined, which may be regarded as the concentration of K close to the mineral surface (Mortland, 1958). The K_{bulk} is the K concentration in the bulk solution. At solution concentrations below K_{crit}, a net release of K occurs, whereas above K_{crit} fixation in the illite interlayers occurs.

MATERIALS AND METHODS

All experimental data were obtained from a monitoring site in the Solling experimental forest, Germany. This site has been part of the International

Biological Program (IBP) since the 1960s (Ellenberg, 1971). The site is located on the Solling plateau, 800 m above sea level. The soil was classified as a spodo-dystric Cambisol and carries a 100 yr old spruce forest (*Picea abies* Karst.). A detailed site description is given by Ellenberg (1971) and Matzner (1989). Our analysis focused on the upper 90 cm of the mineral soil. The upper 60 to 75 cm of the profile consists of loess deposited approximately 10 000 yr ago, followed by a layer of loess mixed with weathered triassic sandstone down to 90-cm depth. The soil fraction < 2 mm contains approximately 10% feldspars and 18% illitic miner-als. Mass fraction (%) of MgO increase from 0.35 at 5-cm depth to 1.03 at 75-cm depth. Mass fractions (%) of K_2O increase from 1.9 at 5-cm depth to 3.24 at 75-cm depth. X-ray analysis (Table 6–1) showed an increase of 1.0 nm illite minerals with depth with a concurrent decrease of vermiculite. The x-ray observations are consis-tent with a simple transformation if 1.0-nm illite to 1.4-nm vermiculite due to weath-ering or replacement of interlayer K. Most likely, chlorite was formed by Al-hydroxy interlayering of illite during soil formation (Tarrah, 1988). Minor amounts of kaoli-nite were found throughout the profile. There was no evidence of smectite presence. Hereafter, the illite/vermiculite/chlorite assemblage will be referred to as illite.

Historical Weathering

The bulk chemistry of six particle size fractions at depth intervals of approx-imately 10 cm was measured by x-ray fluorescence spectrometry (XRD) (Groβkurth, 1993, unpublished data). Samples were air dried, organic matter was removed by oxidation with H_2O_2 at room temperature, and particle size fractions were separated gravimetrically. Unfortunately, the Solling profile contains no C horizon of loess material. A profile 30 km southeast of Solling (location Spanbeck; Tarrah, 1988), however, consists of a >180-cm deep homogeneous loess deposit with a mineralogy similar to that of the Solling site. We assumed that the bulk chemistry of the Spanbeck soil at 180-cm depth is representative of the parent material of the upper 60-cm loess layer at Solling. The ZrO in the 6 to 60-µm particle size fraction, where it was most abundant, was used as an internal standard oxide. As no ZrO content was available for the reference C horizon of the Spanbeck profile, it was calculated as $ZrO_{ref,180cm} = SiO_{2,ref,180cm}/SiO_{2,Solling\ 60cm}$ • $ZrO_{Solling\ 60\ cm}$. Weathering rates were calculated according to Eq. [4], exemplified for K

Table 6–1. Mineralogy of Solling <2 um soil by x-ray powder diffraction.

Depth	Illite[†,‡]	Vermiculite	Chlorite	Kaolinite
cm				
5–10	+	+++	−	+
10–15	+	+++	(+)	+
40–50	++	++	+	+
60–75	+++	+	+	+

[†]Identification criteria: illite, 1.0 nm on Mg saturation; vermiculite, 1.4 nm on Mg saturation, no changes on glycerolation, collapse to 1.0 nm on K saturation at 150°C; chlorite, as two but no col-lapse on K saturation at 150°C; Kaolinite, 0.7 nm on Mg saturation, 0.7 nm on K saturation at 150°C.
[‡]Mineral presence ranges from minor amount, (+) to abundant (+++).

$$R_w = \sum_{x=0cm}^{60cm} \left[\frac{M_{ref.ox, \Delta x}}{M_{ref.ox, rs}} M_{K_2O, rs} - M_{K_2O, \Delta x} \right] / M_{K_2O} 2\rho_{\Delta x} \Delta X / \Delta T 10^3 \quad [4]$$

where:

R_w	=	weathering rate ($kmol_c$ ha^{-1} yr^{-1})
Δx	=	layer of thickness Δx (cm), over which mineral contents were averaged
ΔT	=	assumed time period over which weathering occurred (yr)
ρ	=	bulk density of layer-Δx (g cm^3)
M	=	mass percentage of element oxides
rs	=	reference samples

Weathering from Element Budgets

The Solling data set provides an 18 yr continuous time series of the chemistry of bulk precipitation, throughfall and soil water. For a detailed description of monitoring procedures see Matzner (1989). Weathering rates were estimated from the balance of atmospheric inputs, output with drainage water at 90-cm depth, estimated output by net uptake in biomass, and measured changes of element storage in litter and exchangeable pools over the period 1979 to 1991. Matzner (1989) calculated weathering rates for the period 1973 to 1983 with this method. Here, we discarded the first 6 yr of the observations in Solling, because during this period the pool of exchangeable base cations decreased appreciably (Matzner, 1988), which introduced large uncertainties in the calculation of (low) weathering rates. Weathering was calculated as

$$R_x = - \sum_{1973}^{1991} J_{x,atmosp.} + \sum_{1973}^{1991} J_{x, 90 cm} + \sum_{1973}^{1991} J_{x,net upt.} + \Delta S_{x,litter} - \Delta S_{x, ex} \quad [5]$$

where:

R	=	weathering rate ($kmol_c$ ha^{-1} yr^{-1})
X	=	elements Na, K, Ca, Mg and Al
$J_{x,atmosp}$	=	Atmospheric deposition fluxes, calculated from measured monthly throughfall water fluxes, bulk deposition and estimated dry deposition as described by Matzner (1988, 1989).
$J_{x,90 cm}$	=	Solute drainage fluxes at 90 cm, calculated from measured monthly soil water concentrations and calculated water fluxes (Wesselink et al., 1994).
$J_{x,net upt}$	=	Net uptake of elements in biomass, as given by Matzner (1988).
ΔS_{litter}	=	changes in element storage in litter, estimated from inventories in 1979 and 1983 (Matzner, 1989).
ΔS_{ex}	=	change in pools of exchangeable cations, estimated from inventories in 1973, 1979, 1983 (Matzner, 1988) and 1991.

Weathering from Laboratory Experiments

Background

Experimental methods used to study weathering in the laboratory are shown in Fig. 6–1. The experiments aimed at finding parameters for and testing

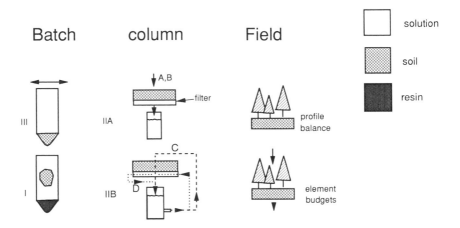

Fig. 6–1. Experimental methods used to measure weathering rates: (i) batch-resin, (ii) column, (A) throughflow and (B) recirculation; (iii) batch equilibrium, and field estimates from bulk chemistry (profile balance) and input-output element budgets (for detailed explanation see text).

of Eq. [2] and [3] that describe mineral weathering. In Experiment 1, a batch-resin experiment, solution chemistry was controlled at conditions strongly undersaturated with respect to all relevant mineral phases. From these experiments the rate constants k in Eq. [2] and [3] could be estimated. Also, at constant and very low concentrations of K_{bulk} (Eq. [3]), the cumulative release of K would show whether the release of K was linear with time (Eq. [3a]) or with the square root of time (Eq. [3b]). To account for the effects of temperature and soil depth on the rate constants, Experiment 1 was carried out at two temperatures, 18 and 5°C, with soil from 10- and 40-cm depth. In Experiment 2A, a throughflow column experiment, the input solution chemistry was manipulated to estimate the pH dependency of weathering and estimate parameter m in Eq. [2]. The flowthrough columns of Experiment 2A were recirculated in Experiment 2B, which forced a continuous increase of reaction products in the system. Experiment 2B was carried out to estimate the parameters Q_{eq} and n in Eq. [2]. The occurrence of a "critical" K concentration (K_{crit}), as given in Eq. [3], was studied separately in Experiment 3. In a batch experiment, nonspecifically bound K was removed, different initial K concentrations were added, and the kinetics of K release or fixation with time were studied. Details on the experimental setup are given below.

Experimental Setup

   Experiment 1—Batch Resin Experiments.   Three grams of air-dried soil were put in a dialysis bag and subsequently in a small vessel with 20 mL of 10^{-3} M HCl. Two grams of H-saturated cation exchange resin [Dowex, Janssen Chimica, Tilburg, the Netherlands, 50–200 mesh (150–300 μm)] were added. The exchange resin controlled pH at 2.75 and accumulated released cations. Cation concentrations in solution remained low; after 4 wk concentrations of K, Mg, and Al in solution were ($2.6 \cdot 10^{-7}$, $1.3 \cdot 10^{-7}$, $1.1 \cdot 10^{-5}$ M) respectively. The aqueous pools of Mg and K were negligible compared to those accumulated on the exchange resin. The first resin replacement after 14 d was discarded, because blank samples (with resin only) still contained adsorbed base cations. Probably, after the first extraction, a large part of exchangeable K and Mg was removed. Feigenbaum and Shainberg (1975) reported in a similar experiment that readily exchangeable K was removed within 2 h. Hereafter, the resins were replaced after 21, 57 and 156 d. Reaction products adsorbed on the resin were retrieved by slow percolation with 25 mL of 2 M HCl. The K and Mg were analyzed by atomic absorption spectrometry (AAS) and Si was determined by colorimetry (blue silica molybdic acid complex). All batch experiments were carried out in duplicate with soil material from 10- and 40-cm depth at two temperatures: 5 and 18°C.

Experiment 2—Column Experiments.   Air-dried soil from 40- to 50-cm depth was packed in soil columns (4.5-cm diam., 1-cm height) which were leached under different percolation conditions.

A.   Flow-through Experiments.   Before starting the actual experiments, the columns were percolated with 60 mL of 0.1 M Ba(Cl)$_2$ solution to remove exchangeable base cations, and for 3 d with HCl (pH 3.0) at a rate of approximately 2 mL h^{-1}. Subsequently, Column A was percolated with HCl (pH 3.2) solution for 700 h followed by a percolation with a $(1 \cdot 10^{-3}M)$ (Al)$_2$(SO$_4$)$_3$ solution (pH 2.82) for 600 h. Column B was percolated with a $(0.4 \cdot 10^{-3}M)$ (Al)$_2$(SO$_4$)$_3$ solution (pH 2.82) for 700 h followed by percolation with a $(1 \cdot 10^{-3}M)$ (Al)$_2$(SO$_4$)$_3$ solution (pH 2.82) for 600 h. Percolation rates were aproximately 1.3 mL h^{-1} in these experiments.

B.   Recirculation Experiments.   In Column C, the initial input solution of HCl (pH 3.2) was recirculated for 700 h. Sampling was done from a recirculation vessel (Fig. 6–1). A similar experiment was carried out with Column D, however, solution was not recirculated through the soil column but through a water reservoir connected to the soil column by a porous filter (Gelman, Versapor-200, van Seenus bv., Almere, the Netherlands; 0.2-μm pore size). Thus transport of reactants and reaction products in the soil column was by diffusion only (Fig. 6–1). In the column experiments, ultrapure water was used. Samples obtained during the experiments were stored at 3°C. The Na, K, Ca and Mg were analyzed by atomic absorption on a Perkin-Elmer graphite tube autoanalyzer system. The Al and Si were determined by colorimetry (measuring Al-pyrocatechol violet respectively blue silica molybdic acid complex).

Experiment 3—Batch Experiments.   The effect of the K concentration in solution on release from or fixation in illite interlayers and the occurrence of a critical K level (K_{crit}) was tested in batch experiments. Potassium in illite can be found "fixed" in interlayers, specifically adsorbed on so-called "wedge-sites" or nonspecifically adsorbed on planar (external surface) sites (Fanning & Keramidas, 1977). Twelve samples of 2.5 g of soil from 10- and from 40-cm depth were washed with three subsequent aliquots of 20 mL of 0.1 M Ba(Cl)$_2$, to remove nonspecifically adsorbed K. Next, 20 mL batches of 0, 5 10, 30, 100 and

$300 \cdot 10^{-6}$ M KCl in a 0.01 M Ba(Cl)$_2$ background were added to duplicate soil samples. Samples were gently shaken at room temperature. After 0.5, 2.5, 24, and 300 h, samples were centrifuged and K activity was measured with an Orion Model 93-19 K selective electrode (Orion, Breda, the Netherlands). A linear calibarion curve was obtained for the experimental concentrations range between $3.2 \cdot 10^{-6}$ and $0.3 \cdot 10^{-3}$ M of K.

Results of the laboratory experiments were used to estimate the parameters of the proposed rate Eq. [2] and [3] by means of nonlinear regression (SPSS). The k" in Eq. [3b] was estimated from the cumulative K release in Experiment 1 (batch-resin) and 2A (column flow-through). Parameters a, b, and Q_{eq} in Eq. [2] were estimated from Experiment 2B (column recirculation). Parameters k and m in Eq. [2] were estimated from Experiments 2A (column flow-through) as well as from Experiments 1, 2A, and 2B. As estimation of parameter n in Eq. [2] proved difficult, all experiments were used in its estimation (results are given in Table 6–3).

RESULTS AND DISCUSSION

Field

Historical and present field weathering rates calculated from Eq. [4] and [5] are given in Table 6–2. The calculation of weathering rates from element budgets involved uncertainties in all fluxes considered. Yet, positive weathering fluxes

Table 6–2(a).   Present weathering rates in the 0 to 90 cm Solling soil obtained from element budgets[†] in the Solling spruce (genus *Picea*) site for the period 1979 to 1991.

	Mg	K	Ca	Na	Al
	kmol$_c$ ha^{-1}				
$\sum$atmospheric input	3.1	2.4	10.3	8.5	3.2
$\sum$drainage output	-5.8	-1.1	-7.7	-10.4	-91.2
$\sum$net uptake	-1.1	-1.9	-3.9	‡	‡
Δlitter	-1.5	-1.0	-3.9	‡	‡
Δexchange	1.3	1.0	4.4	1.3	-8.0§
$\sum$weathering	4.0	0.8	0.8	0.6	96
	(kmol$_c$ ha^{-1} yr^{-1})				
Weathering	0.33	0.06	0.06	0.05	8.0

†Positive signs indicate sources, negative signs indicate sinks.
‡assumed to be negligible
§assumed to equal the sum of exchangeable base cation loss

Table 6–2(b).   Historical weathering rates in the 0 to 60 cm Solling soil over the past 10 000 yr and molar Mg K Al Si ratios of weathering.

	Si	Mg	K	Ca	Na	Al
	(kmol$_c$ ha^{-1} hr^{-1})					
Weathering	1.46	0.30	0.20	0.083	0.077	1.60
Molar ratios	3.5	0.36	0.47			1.3

were calculated which were in the same order as the independently derived historical rates. The estimates of present weathering rates of Mg and K compare well with those given by Matzner (1989) calculated for the period 1973 to 1983 (0.36 and 0.08 $kmol_c ha^{-1} yr^{-1}$ respectively). Contrary to the 1973 to 1983 period (Matzner, 1989), positive weathering rates were calculated for Ca and Na in the 1979 to 1991 period.

Weathering rates of Ca, Na and K release for the past 10 000 yr and for the present can be considered low. For Ca, the annual release is even lower than the net annual storage of Ca in biomass (Matzner, 1989), which emphasizes the role of atmospheric deposition as a source of Ca to the forest. As K release from microcline feldspar is generally considered to be slower than Na release from albite (e.g., Sverdrup, 1990), the estimated K release can be largely attributed to the weathering of illite. All Mg release is presumably from clay mineral dissolution, as the Solling soil contains no other Mg-bearing minerals. Note, that the molar K Mg Al Si ratio of illite in the C horizon of the reference profile was estimated at 0.5:0.29:2.0:3.5 (Tarrah, 1989), whereas long-term weathering at Solling occurred at a ratio of 0.47:0.36:1.3:3.5 (see Table 6–2). Therefore, dissolution of illite appears to be nonstoichiometric and preferentially for Mg (Tarrah, 1989).

Interpretation of differences between historical and present rates fo element release is difficult, because both pH and mineralogy will have changed considerably during the past 10 000 yr. Present rates may be lower than historical rates due to depletion of mineral pools, which seems the case for K. On the other hand, the Solling soils receive high anthropogenic acid depositions (Matzner, 1988, 1989), which may have enhanced mineral weathering, as is probably the case for Mg. Enhanced weathering due to anthropogenic acidification obviously explains the present high rate of Al release, which is six times higher than the average over the past 10 000 yr.

LABORATORY

Experiment 1+2 (Batch Resin and Flow-through)

The cumulative release of K and Mg in the batch-resin experiments is shown in Fig. 6–2. Linear release curves were obtained for Mg, except for the 40-cm/18°C samples, where the release seems to be curvilinear with time. For K, parabolic release curves were obtained, except for the 10-cm/18°C samples. A pronounced effect of sampling depth was found on release rates of Mg. In the experiment at 5°C, the release from the 40-cm samples was 1.5 × the release from the 10 cm samples, which compared well with the gradient of magnesium oxide with depth as obtained from the XRF analysis [mass fractions (%) of 0.38 for MgO at 10 cm and 0.61 for MgO at 40 cm]. The dependency of K release on soil depth was minor at 18°C and it changed with time for the 5°C experiment. The low dependency of K release on soil depth also is evident from the K_2O contents (XRF). Mass fractions (%) of K_2O were measured at 2.4- and 40-cm depth and 1.97 at 10-cm depth. Figure 6–2 shows a higher release rate of K and Mg in the 18°C experiments. A measure for the effect of temperature on reaction rate is given by the Arrhenius activation energy, E_a (e.g., Sparks, 1988). For Mg, E_a was calculated at ±61 $kJ mol^{-1}$ in the 10-cm samples and ±76 $kJ mol^{-1}$ in the 40-cm samples.

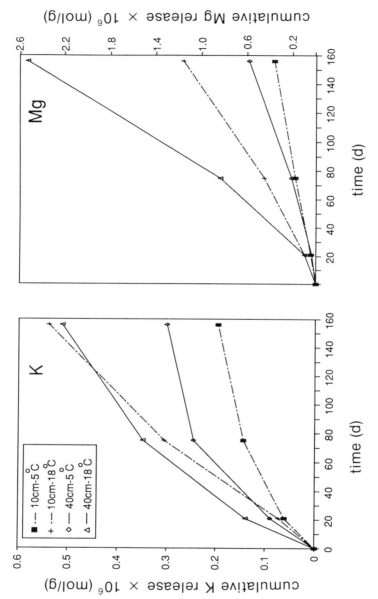

Fig. 6–2. Cumulative release of K and Mg in batch resin experiments. Samples from 10 cm at 5°C, 10 cm/18°C (+), 40 cm/5°C (◆) and 40 cm/18°C (△) (averages of duplicate measurements are shown).

For release of K, E_a was calculated at ± 38 kJ mol⁻¹ for the 10-cm samples and ± 18 kJ mol⁻¹ for the 40-cm samples during the first 57 d.

Figure 6–3 shows the cumulative release of Mg, K, Al and Si from the flowthrough column (Experiment 2A) and observation points of the batch resin Experiment 1. Whereas the release of Mg responded strongly on pH, releases of K seemed unaffected.

Obviously, dynamics of K and Mg release differed strongly in Experiments 1 and 2. The parabolic release curves of K are indicative for diffusion controlled release as described by Eq. [3] (Luce et al., 1972). Activation energies of K release were <40 kJ mol⁻¹, which is also indicative for diffusion controlled release (Sparks, 1988). Mortland and Ellis (1959) report an E_a of 15 kJ mol⁻¹ for diffusion controlled release of K from vermiculite. Activation energy for Mg release was >40 kJ mol⁻¹ indicating chemically controlled release rates (Sparks, 1988). The different nature of Mg and K release also appears from their dependency on pH. The effects of pH on K release from illite has been discussed extensively in the literature. Our results confirm those of Wells and Norrish (1968), who concluded that pH was of minor importance in the replacement of interlayer K. In a recent paper, Simard et al. (1992) studied K and Mg release in mica containing soils. They found no relationship between the amount of K and Mg released with citric acid. In contrast to our experiments, a parabolic release curve was reported also for Mg.

Experiment 2B (Column Recirculation)

A general rate equation for chemically controlled release of Mg from illite was given by Eq. [2]. In Experiment 2B the effect of the equilibrium term in Eq. [2] (IAP/Q_{eq}) on release of Mg was investigated. Solutions were recirculated (Fig. 6–1, Column C+D) and reaction products were allowed to accumulate. Typically, concentrations of Mg, Al, Si and pH remained constant after 400 h. Different equilibrium concentrations for Mg, Si, Al and pH were reached on Columns C and D, after 400 h. These equilibrium concentrations were used to estimate the stoichiometry and equilibrium solution activity product of the illitic phase controlling Mg release (Fig. 6–4). Parameters a and b give the reaction stoichiometry for the dissolution of a Mg-Al-Si illitic phase (Eq. [2a]). Previously, the molar Mg Al Si ratios of illite was estimated at 0.14:2.4:3.3 in the 0 to 40 cm soil of our reference profile (Tarrah, 1989). The molar ratios of dissolution obtained from Experiment 2B (0.15:1.03:3.5) therefore indicate nonstoichiometric dissolution of illite. Nonstoichiometric weathering also was apparent from the long-term, historic, weathering rates obtained for Solling (see FIELD discussion).

Experiment 3 (Batch)

A general rate equation for diffusion controlled release of K from illite was given by Eq. [3]. Fig. 6–5 shows the K concentrations in solution and the amounts of K released to, or retrieved from solution in the batch Experiment 3 after 2.5, 24 and 300 h. During the first 24 h, kinetics of K release and fixation were fast compared to the batch-resin Experiment 1, probably due to adsorption or desorption on selective "wedge" sites (Fanning & Keramidas, 1977). From

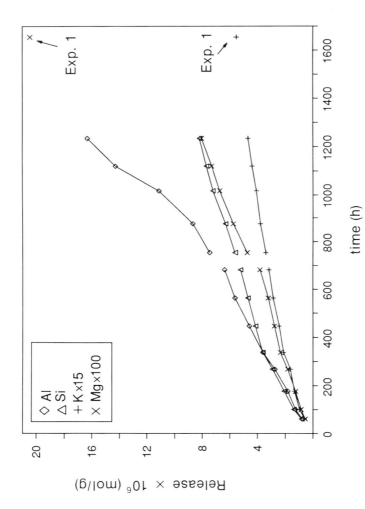

Fig. 6–3.   Cumulative release of K (+), Mg (×), Al (◆) and Si (△) in flowthrough column experiments, calculated as column output minus column input: 0 to 700 h, input solution HCl (pH 3.2), output pH ≈3.8; 700 to 1300 h, input solution (1 • 10⁻³ M) Al₂(SO4)₃ (pH 2.8), output pH ≈3.35. For K and Mg, releases from Experiment 1 (pH 2.75) also are indicated.

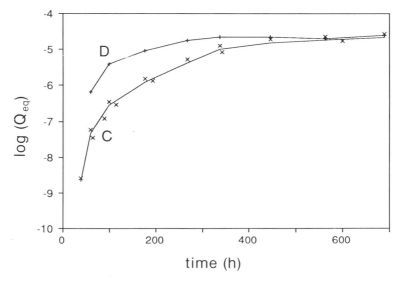

Fig. 6–4. Activity products of hypothetical illitic phase: $Q = \{Mg\}^a\{Al\}^b\{Si\}^{3.5}$, where $\{\}$ indicate solution activities. The a and b were estimated at 0.15 and 1.04 respectively by minimizing the differences in Q values from Column C and D after 400 h.

24 h on, the release slowed down, most likely due to diffusion controlled release or fixation of interlayer K. A "critical" K concentration is apparent from Fig. 6–5, which seems to confirm the validity of the proposed Eq [3b] for K release. The K_{crit} was estimated at $17 \cdot 10^{-6}$ M for the 40-cm samples and $27 \cdot 10^{-6}$ M for the 10-cm samples. At bulk solution concentrations below K_{crit}, K was released to solution, whereas K was retrieved from solution when the K concentration in the bulk solution was higher than K_{crit}.

MODEL VALIDATION

Eq. [3b] was tested by using it to predict the slow kinetics of K release in batch Experiment 3 between 2.5 and 300 h. Rate constant k^{11} in Eq. [3b] was estimated from Experiment 1 and 2A (Table 6–3) and K_{crit} values from Experiment 5. Concentrations at 2.5 h (Fig. 6–5) were used as initial values in Eq. [3b]. Predictions and measurements at 24 and 300 h are shown in Fig. 6–5. Predictions seem to compare well with observations, except for the highest K concentrations ($>50 \cdot 10^{-6}$ M) in the 40-cm samples. Bulk solution concentrations of K exceeding $50 \cdot 10^{-6}$ M are, however, not representative for soil solutions in Solling, where K concentrations are in the range of 10 to $40 \cdot 10^{-6}$ M.

Parameter values $k = 2 \cdot 10^{-5}$ and $m = 0.69$ (Table 6–3), obtained from the flow-through column Experiments 2A, were used to predict release of Mg in the batch-resin experiment (40 cm/18°C). The calculated release, however, underestimated the observations by a factor of three. Probably, release of Mg in the batch-resin experiments was enhanced compared to the column experiment, due to much

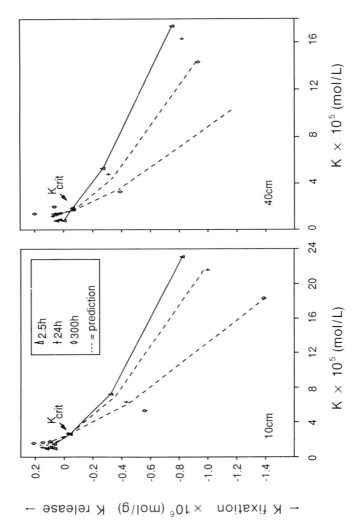

Fig. 6–5. K kinetics in batch-equilibrium experiment. Shown is the quantity of K adsorbed (−) or released (+) by the Ba-saturated soil, against K activity measured in the bulk solution. Initial concentrations were: 0, 5, 10, 30, 100 and $300 \cdot 10^{-6}$ M KCl in a 0.01 M Ba(Cl)$_2$ background. Averages of duplicate measurements and predicted K dynamics (see text) are shown.

Table 6–3.   Estimate parameters for Mg and K release according to Eq. [2] and [3b], and standard error of estimates (SD) of model fit (R^2), number of observations (n) and experiments from which observations were obtained.

Eq. [2]	Estimated	SD	R^2	n	Experiment
$\log(Q_{eq})$	−4.54	0.99	0.98	10	2B
a	1.03	0.092	0.98	10	2B
b	0.15	0.51	0.98	10	2B
n	0.28	0.42	0.58	28	1, 2A, 2B
			mol kg^{-1}h^{-1}		
k	20.3 • 10^{-6}	60.2 • 10^{-6}	0.39	14	2A
m	0.69	0.37	0.39	14	2A
			mol kg^{-1}h^{-1}		
k	0.21 • 10^{-3}	0.30 • 10^{-3}	0.58	28	1, 2A, 2B
m	0.99	0.22	0.58	28	1, 2A, 2B
Eq. [3b]					
			L kg^{-1} h$^{-1/2}$		
k''	0.435	0.025	0.89	13	1, 2A

†Parameters refer to experiments at 18°C with soil material from 40- to 50-cm depth.

lower Al-concentrations in the bulk solution. Sverdrup (1990) reported an increase of mineral weathering with decreasing Al concentrations. Also, dissolution of protective Al-coatings, or Al-interlayers, as reported by van Grinsven (1988) may have enhanced Mg release. Because we had little certainty about the actual cause of the differences in Mg release in Experiment 1 and 2, a second estimate of k and m was made, which included results from all experiments (Table 6–3).

Note, that estimates of k, b and n (Eq. [2]) were not significant (Table 6–3), which most probably results from an overparameterization of Eq. [2], but also may result from an inappropriate model structure (i.e., Eq. [2]). Therefore, independent experiments to further test the validity of Eq. [2] are needed.

EXTRAPOLATION TO THE FIELD

To extrapolate laboratory observations to the field, Eq. [2] and [3b] were applied to field conditions. The predicted field rates were compared with the measured field rates shown in Table 6–2. Weathering rates were predicted for the upper 60 cm of the Solling soil. The predictions were carried out stepwise (Table 6–4) and started with the rates derived from experiment 1 (pH 2.8, 18°C). Next it was corrected for:

Soil Depth.   From the batch-resin experiments with soil from 10- and 40-cm depth, it appeared that release of Mg was linearly related to contents of MgO. Therefore, a linear relation between rate constant k in Eq. [2] and MgO contents was assumed. For K release the dependency on depth of sampling was less clear. For the purpose of extrapolating to field conditions, it was assumed that k^{11} in Eq. [3b] was independent of depth. Detailed data on bulk density were available (Groβkurth, 1993, unpublished data).

Table 6–4. Stepwise extrapolation of rates of Mg and K weathering inferred from laboratory experiments to field conditions.

	Mg	K
——————— (kmol$_c$ ha^{-1} yr^{-1}) for 0–60 cm ———————		
Experiment 1	85	10.5
Stepwise corrections		
soil depth	44	10.5
+		
tempetature	11	5.6
+		
pH	0.58†	–5.6
+		
IAP/K_{bulk}	‡	0.25
Field estimates	0.23–0.33	0.06–0.14

†See Table 6–3. Values used are $k = 0.21 \cdot 10^{-3}$ and $m = 0.99$, for $k = 20.3 \cdot 10^{-6}$ and $m = 0.69$, a field estimate of 0.93 kmol$_c$ ha^{-1} yr^{-1} is obtained after correction for soil depth, temperature and pH.
‡Not estimated, see discussion.

Temperature. To correct k and k'' in Eq. [2] and [3b] for temperature, activation energies as derived in Experiment 1 were used: 61 kJ mol^{-1} for Mg release in the 0 to 30 cm soil and 76 kJ mol^{-1} for Mg release in the 30 to 60 cm soil, 38 kJ mol^{-1} for K release in the 0 to 30 cm soil and 15 kJ mol^{-1} for K release in the 30 to 60 soil. The average soil temperature was assumed to match the long-term average air temperature in Solling at 6.5°C.

Soil Solution Chemistry. Soil solution chemistry is measured on a routine basis in Solling. Monthly values of chemistry at 10-, 20-, 40- and 90-cm depth were available for the 1982 to 1991 period (Matzner, 1988). Monthly solution chemistry at 60-cm depth was calculated as the average of 40 and 90 cm observations. The K release for the 1982 to 1991 period was calculated by averaging monthly predicted rates to annual rates. Similarly Mg release at field pH was calcualted. The correction for IAP/Q_{eq} at field conditions will be discussed separately.

Table 6–4 shows predictions of field weathering rates, subsequently corrected for soil depth, temperature, and solution chemistry at field conditions. The final K release under field conditions between 1982 and 1991 was predicted at 0.25 kmol$_c$ ha^{-1} yr^{-1}, which is close to the field estimates. The term (K_{crit}–K_{act}) has a strong effect on the release rate of K, because concentrations of K in the field were close to the critical K concentration (K_{crit}) that we measured in the laboratory. Note, however, that on a monthly scale, predictions fluctuated considerably [maximum monthly release (+) or fixation (–) rates of 0.15 kmol ha^{-1} were calculated] as a result of deviations of soil solution K concentrations from K_{crit}. Over longer periods, the monthly predictions balanced to an average release of 0.25 kmol ha^{-1} yr^{-1}. The rather high predictions of monthly K release or K fixation at field conditions may be explained by an effective diffusion constant (k'' in Eq. [3b]) in the field, that is lower than the value used in this study. Unsaturated moisture conditions and soil structure may increase the transport distance from the mineral surface to the bulk solution and therefore decrease the effective diffusion constant. Diffusion of K to the bulk solution can be considered as two sequential steps, (i) diffusion from the illite interlayer and (ii) diffusion on the interparticle scale. The slowest of the two steps will determine the rate of K release to the bulk solution. In Experiment 2, the second diffusion step was eliminated due to a sufficiently high convective mass transport. In Experiment 1, transport was by diffusion only, over a diffusion distance of approximately 2 cm. Still, similar release rates of K were observed in Experiments 1 and 2 (see Fig. 6–3), which indicates that the largest resistance for K transport was interlayer diffusion. Therefore, we expect a minor effect of unsaturated moisture conditions and soil structure on K release at field conditions as long as these conditions do not effect interlayer diffusion. Indeed, Haagsma and Miller (1963) reported a small effect of moisture content on K release in soil-resin mixtures at 5°C.

Predictions of Mg-release are most strongly affected by temperature and pH. After correction for pH and temperature, the extrapolations of laboratory data to field conditions were 2 to 4 times higher than the field estimates. The subsequent correction for mineral saturation, the $(1 - [Q/IAP]^n)$ term in Eq. [2], was not possible at field conditions, as, unfortunately, the routine monitoring of soil water chemistry in Solling did not include the measurement of Si. However, in an adjacent plot where monitoring started in 1991, 20 samples from 40- and 70-cm depth respectively were obtained which included Si measurements. The IAP with respect to the hypothetical $Mg_{0.15}Al_{1.04}Si_{3.5}$ phase (Exp. 2B) was calculated for these samples and plotted in a stability diagram shown in Fig. 6–6, together with the stability lines of gibbsite, kaolinite, quartz and the hypothetical Mg-Al-Si phase proposed in this study. Also, the temperature dependency of the quartz, kaolinite and gibbsite lines are shown. From Fig. 6–6 it appears that soil water samples fall within the stability fields of gibbsite and quartz and also are close to the stability line of the hypothetical Mg-Al-Si mineral. Part of the 40-cm samples is oversaturated with respect to the hypothetical mineral phase which, in terms of Eq. [2], means that Mg tend to precipitate from solution. However, it also appears from Fig. 6–6 that the effect of temperature on mineral stability is distinct and can probably not be neglected for the hypothetical Mg-Al-Si phase. Unfortunately, the effect of temperature on the stability of the hypothetical Mg-Si-Al mineral, which was established at 18°C, is unknown.

CONCLUSIONS

Weathering of Mg and K from an illite-containing acid forest soil in Solling, Germany, was investigated at field and laboratory conditions. Factors that were expected to control weathering in the field were varied at laboratory conditions. Mineral depletion, which was correlated with soil depth, temperature and solution chemistry proved to be important factors in controlling release rates of Mg and K from illite. Potassium release from illite was controlled by interlayer diffusion, whereas the release of Mg was chemically controlled. The proposed diffusion model for K release predicted high potential rates of K release at field conditions. However, predicted actual field rates were much lower, and close to field estimates, because K diffusion from interlayer space was limited by K concentrations in soil water, which were close to threshold values for interlayer release, obtained in the laboratory. Potential release rates of Mg at field conditions were a factor 2 to 4 higher than field estimates, after correction for temperature and pH.

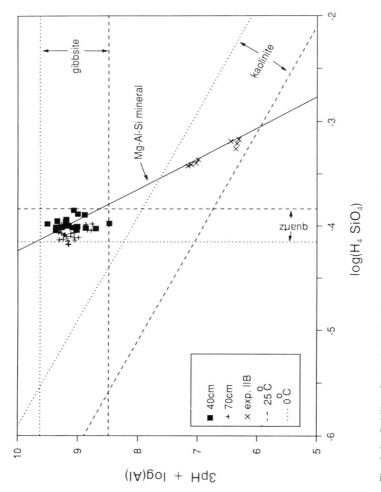

Fig. 6–6.   Stability of several aluminosilicates. Stability constants of quartz, kaolinite and gibbsite at 25°C are given by Robie et al. (1978). Stability constants at 0°C were calculated from constants at 25°C and standard enthalpys of formation given by Robie et al. (1978). For the "hypothetical" Mg-Al-Si phase and the soil samples, equilibrium with $5 \cdot 10^{-5}$ M Mg^{2+} was assumed.

The remaining discrepancy between estimated and observed field rates of Mg release may be caused by underestimation of the effect of saturation with the hypothesized reference mineral. To resolve this problem, the stability of the hypothetical Mg-Al-Si mineral needs further study.

The present study illustrates the potential of laboratory experiments to explain the apparent discrepancy between laboratory and field rates by differences in reaction conditions. Still, Mg release was a factor 2 to 4 times higher than observed at field conditions, which will cause a gross underestimation of the vulnerability of the forest ecosystems to external disturbances like acid rain. Therefore, present methods to obtain field weathering rates, i.e., from elemental analysis of the profile (historical weathering) and from input-output budgets (present weathering) remain necessary.

ACKNOWLEDGMENTS

The authors wish to thank Piet Otten, Eef Velthorst, Neel Nakken and Jan van Doesburg for carrying out laboratory analyses and Norbert Lammersdorf for providing the data on Si concentrations in Solling soil water. Discussions with Ed Meijer on interpretation of the data were highly appreciated. We are grateful to Professor Nico van Breemen for critically reviewing this paper. Part of this study was financed by the German Ministry of Research and Technology, Project no. OEF-2019-3.

REFERENCES

Aagard, P., and H.C. Helgeson. 1982. Thermodynamic and kinetic constraints on reaction rates among minerals and aqueous solutions. I. Theoretical considerations. Am. J. Sci. 282: 237–285.

Anbeek, C. 1992. Surface roughness of mineral and implications for dissolution studies. Geochim. Cosmochim. Acta 56:1461–1469.

Ellenberg, H. (ed.). 1971. Integrated experimental ecology. Methods and results of the ecosystem research in the German Solling project. Ecol. Stud. 2. Springer Verlag, New York.

Fanning, D.S., and V.Z. Keramidas. 1977. Micas. p. 195–258. In J.B. Dixon, and S.B. Weed (ed.) Minerals in soils environments. SSSA, Madison, WI.

Feigenbaum, S., and I. Shainberg. 1975. Dissolution of illite—A possible mechanism of potassium release. Soil Sci. Soc. Am. J. 39:985–990.

Flehmig, W., H. Fölster, and J. Tarrah. 1990. Stoffbilanzierung in einer Pseudogley-Parabraunerde aus Löβ unter Anwendung der IR-Phasenanalyse. Z. Pflanzenernährung. Bodenk. 153:149–155.

Haagsma, T., and M.H. Miller. 1963. The release of nonexchangeable soil potassium to cation-exchange resins as influenced by temperature, moisture and exchanging ion. Soil Sci. Soc. Am. Proc. 34:153–156.

Holdren, G.R., and R.A. Berner. 1979. Mechanisms of feldspar weathering-I. Experimental studies. Geochim. Cosmochim. Acta 43:1161–1171.

Jenny, H. 1941. Factors of soil formation: A system of quantitative pedology. McGraw Hill Book Co., New York.

Kirchner, J.W. 1992. Heterogeneous geochemistry of catchment acidification. Geochim. Cosmochim. Acta 56:2311–2327.

Luce, W.L., R.W. Bartlett, and G.A. Parks. 1972. Dissolution kinetics of magnesium silicates. Geochim. Cosmochim. Acta 36:35–50.

Manley, E.P., and L.J. Evans. 1986. Dissolution of feldspars by low-molecular-weight aliphatic and aromatic acids. Soil Sci. 141:106–112.

Matzner, E. 1988. Der Stoffumsatz zweier Waldökosysteme im Solling. Ber. Des Forschungszentrum Waldökosysteme Göttingen, Reihe A. Bd. 40.

Matzner, E. 1989. Acidic precipitation: Case Study Solling. p. 39–84. *In* D.C. Adriano, and M. Havas (ed.) Advances in environmental science. Vol. 1. Springer, New York.

Mortland, M.M. 1958. Kinetics of potassium release from biotite. Soil Sci. Soc. Am. Proc. 22:503–508.

Mortland, M.M., and B. Ellis. 1959. Release of potassium as a diffusion controlled process. Soil Sci. Soc. Am. Proc. 23:363–364.

Nagy, K.L., A.E. Blum, and A.C. Lasaga. 1991. Dissolution and precipitation kinetics of kaolinite at 80°C and pH3: The dependence on solution saturation state. Am. J. Sci. 291:649–686.

Olsson, M., and P.A. Melkerud. 1991. Determination of weathering rates based on geochemical properties of the soil. p. 69–78. *In* E. Pulkkinene (ed.) Environmental geochemistry in northern Europe. Geol. Surv. of Finland. Spec. Pap. 9. Geologian Tutkimuskeskus, Espo, Finland.

Robie, R.A., B.S. Hemmingway, and J.R. Fisher. 1978. Thermodynamic properties of minerals and related substances at 298°K and 1 bar (10^5 Pascals) pressure and at higher temperatures. U.S. Gov. Print. Office, Washington, DC. U.S. Geol. Surv. Bull. 1452.

Schnoor, J.L. 1990. Kinetics of chemical weathering: A comparison of laboratory and field weathering rates. p. 475–504. *In* W. Stumm (ed.) Aquatic chemical kinetics. ES&T Monog. New York.

Simard, R.R., C.R. De Kimpe, and J. Zizka. 1992. Release of potassium and magnesium from soil fractions and its kinetics. Soil Sci. Soc. Am. J. 56:1421–1428.

Sverdrup, H.U. 1990. The kinetics of base cation release due to chemical weathering. Lund Univ. Press, Sweden.

Sparks, D.L. 1988. Kinetics of soil chemical processes. Acad. Press, New York.

Tarrah, J. 1988. Verwitterungsbilanzen von Böden auf der Basis modaler Mineralbestände. Ber. des Forschungszentrums Waldökosysteme, Göttingen. Reihe A, Bd. 52.

van Grinsven, J.J.M. 1988. Impact of acid atmospheric deposition on soils: Quantification of chemical and hydrological processes. Ph.D. diss. Agricultural Univ. of Wageningen, the Netherlands.

van Grinsven, J.J.M., G.D.R. Kloeg, and W.H. van Riemsdijk. 1986. Kinetics and mechanisms of mineral dissolution in a soil at pH values below 4. Water Air Soil Pollut. 31:981–990.

van Grinsven, J.J.M., and W.H. van Riemsdijk. 1992. Evaluation of batch and column techniques to measure weathering rates in soils. Geoderma 52:41–57.

Wels, C.B., and K. Norrish. 1968. Accelerated rates of release of interlayer potassium from micas. p. 683–694. Int. Congr. Soil Sci., Trans. 9th (Adelaide, Australia) 2:683–694.

Wesselink, L.G., J. Mulder, and E. Matzner. 1994. Modeling seasonal and long-term dynamics of anions in an acid forest soil, Solling, Germany Geoderma. (In press.)

7   Developing and Adapting Soil Process Submodels for Use in the Pedodynamic Orthod Model

Marcel R. Hoosbeek and Ray B. Bryant

Cornell University
Ithaca, New York

Many different types of models have been utilized in the history of pedology. In a recent paper Hoosbeek and Bryant (1992) presented a framework for the classification of pedogenetic models based on relative degree of computation, complexity, and level of organization (Fig. 7–1). The first characteristic, relative degree of computation, distinguishes between qualitative and quantitative models. The second characteristic, complexity of the structure used in the model, distinguishes between functional and mechanistic models. The third characteristic, organizational hierarchy, describes the level at which a model aims to simulate a natural system. The pedon was placed at the central i-level. Positive i-levels include the polypedon ($i+1$), catena or catchment ($i+2$), and the soil region ($i+3$). Negative i-levels comprise of the horizon ($i-1$), peds and aggregates ($i-2$), and molecular interaction ($i-3$).

The present paradigms of pedology are based on qualitative models, e.g., Jenny's (1941) *Factors of Soil Formation,* was positioned in the framework as a *qualitative-functional-positive* i-*level* model, and Simonson's (1959) *Generalized Theory of Soil Genesis,* which is primarily concerned with the differentiation of horizons in a profile, was positioned as a *qualitative-functional-negative* i-*level* model. Because data for these qualitative models are usually obtained from the soil landscape at one time, these static observations provide a "snapshot" of the soil system.

This paper examines the development and structure of the ORTHOD computer model which simulates soil forming processes involved in the formation of a Typic Haplorthod. A pedogenetic study was designed to quantify processes over short time increments reflecting seasonal fluctuations in temporal properties. We coin the word "pedodynamics" and define it as "The quantitative integrated simulation of physical, chemical, and biological soil processes acting over short time increments in response to environmental factors." Newly developed and adapted soil process submodels used as building blocks in an integrated pedodynamic model are presented.

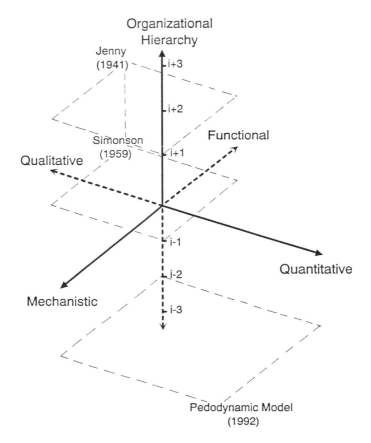

Fig. 7–1.   Classification of pedological models based on relative degree of computation, complexity
and level of organization.

To obtain data for the development and validation of various submodels of the
pedodynamic model, three detailed monitoring sites were installed in a boreal forest in the Adirondack Mountains near Tupper Lake, New York. The soil is a well
expressed Typic Haplorthod developed in sandy glacial outwash deposits and has
never been disturbed by logging or agriculture. The site is nearly level, and there
is no measurable runoff or runon. Tensiometers, soil solution samplers, and ther-
mistors were installed in the Oi, Oa, E, Bh/Bhs, Bs, BC, and C horizons. Redox
potentials were measured in the Oi, Oa, and E horizons. Soil solutions, precipitation, and canopy throughfall samples were collected every 4 wk over 2 yr and were
chemically analyzed for organic and inorganic components.

Each submodel within the pedodynamic model (Fig. 7–2) will be described
separately in subsequent sections of this paper. Since the submodels dealing with
water movement, solute movement, soil temperature profiles, and chemical speciation and precipitation, have been described by other authors, they are

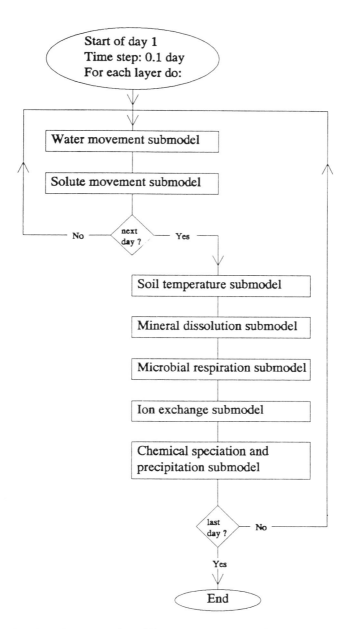

Fig. 7–2. Overview of the submodels within the pedodynamic ORTHOD model. The water and solute movement submodels are called every 0.1d, the other submodels are called daily.

described here briefly. However, the submodels dealing with mineral dissolution, microbial decomposition of organic matter, and ion exchange, are newly developed models resulting from controlled laboratory experiments and are discussed in some detail. Taken together, the submodels, as a whole, represent the complex processes that govern the chemistry and movement of Al, silica, organic components, and other chemical species in Spodosols. The ORTHOD model was applied to the simulation of C dynamics and was used to test several hypotheses on Spodosol formation. However, this paper focuses only on development of the individual submodels. The integration of the submodels, the field data, the validation of the pedodynamic model, and the pedogenetic implications, are beyond the scope of this chapter.

WATER MOVEMENT

The movement of water in the soil is fundamental to any quantitative mechanistic pedogenetic model since water plays a major role in transfer and transformation processes in soils. Realistic representation of saturated and unsaturated flow in a typically nonisotropic medium like soil is a complex problem. Several deterministic models have been developed that simulate the flow of water in the unsaturated zone based on Darcy's law and the continuity principle.

$$ q = -K \frac{\delta H}{\delta z} \left(cmd^{-1} \right) \qquad \frac{\delta \Theta}{\delta t} = - \frac{\delta q}{\delta z} \left(d^{-1} \right) $$

Combination yields a partial differential equation in terms of hydraulic head, called the Richard's equation

$$ \frac{\delta \Theta}{\delta t} = \frac{\delta}{\delta z} \left[K \frac{\delta H}{\delta z} \right] $$

Using the pressure head form of the flow equation and definition of the differential moisture capacity C, $C = d\Theta/d\psi$ yields the flow equation for predicting water movement in layered soils

$$ \frac{\delta \psi}{\delta t} = \frac{1}{C(\psi)} \frac{\delta}{\delta z} \left[K(\psi) \left(\frac{\delta \psi}{\delta z} - 1 \right) \right] $$

The sandy soil at the study site permitted the use of a model based on the Richard's equation. The LEACHM model (Wagenet & Hutson, 1987) uses a numerical solution to the Richard's equation and was selected for use in the ORTHOD model.

Real-time weather data were obtained from the nearby Tupper Lake Sunmount weather station. Undisturbed core samples of each horizon were used to determine bulk densities and water retention curves. The water retention curves were established by equilibrating the undisturbed cores on pressure plates at pressure ranging from 10 to 1500 kPa. Measurement of the water content at each pressure resulted in θ-h (h = pressure head of soil water, θ = volume fraction of liquid) relations for each horizon. The RETFIT program (Hutson & Cass, 1987) was used to fit the retention data and to calculate the "a" and "b" parameters of

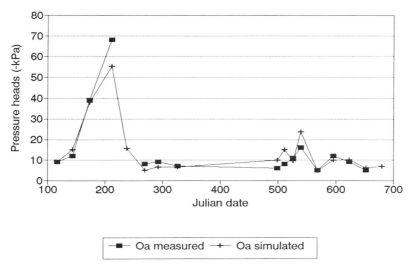

Fig. 7–3.   Measured and simulated pressure heads (–*k*Pa) of the Oa horizon, central site.

Campbell's equation (Campbell, 1985) for use in LEACHM. With the θ-*h* relations defined for each horizon, LEACHM was calibrated by adjusting the saturated hydraulic conductivities of each horizon. In the LEACHM and ORTHOD models the soil profile is divided into layers of equal thickness, 25 mm, whose properties are assumed to be concentrated at the nodes (a horizon may have several layers). Measured and simulated tensiometer values of the Oa horizon are presented in Fig. 7–3. Simulated values gave a good fit for measured values. Given precipitation and evaporation input, this submodel gave a relatively good representation of soil moisture conditions in all horizons at any point in time and over the full range of soil moisture conditions.

SOLUTE MOVEMENT

The LEACHM model (Wagenet & Hutson, 1987) uses a numerical solution to the convection-dispersion equation to estimate chemical fluxes. Solute movement consists of three components: convective, diffusive, and dispersive transport. Combining the equations of these three processes leads to the following expression for solute flux (J_s)

$$J_s = -\theta D\left(\delta C / \delta x\right) + qC \qquad\qquad \text{(van Genuchten \& Wierenga, 1986)}$$

where θ is the volumetric water content, C is the volume averaged solute concentration, x is distance, q is the volumetric fluid flux density, and D is the summation of the molecular diffusion and mechanical dispersion coefficients. Both the diffusion and dispersion coefficients were assumed to be negligible for a sandy soil with a relative large downward convective flow throughout the year.

The solute movement between two layers is then dominantly convective flow, which is the product of the volumetric fluid flux density and the volume averaged solute concentration (Wagenet, 1986)

$$J_s = qC$$

This simplification significantly reduced computation time and did not cause significant error when using relatively thin layers (25 mm) in the simulations.

SOIL TEMPERATURE

Heat transfer and soil profile temperatures were simulated with LEACHM as described by Tillotsen et al. (1980). The heat flow equation is

$$\rho C_p \delta T / \delta t = \delta \left(K_t(\theta) \delta t / \delta z \right) / \delta z$$

where ρ is bulk density (kg•m^{-3}), C_p is gravimetric heat capacity of the soil (J•m^{-3}•°C^{-1}), T is temperature (°C), t is time (s), z is depth (m) and $K_t(\theta)$ is the thermal conductivity of the soil (J•m^{-1}•s^{-1}•°C^{-1}) at water content θ (m^3•m^{-3}). The volumetric heat capacity is calculated from

$$\rho C_p = \rho_s C_s + \theta \rho_w C_w$$

where ρ_s and C_s are the bulk density and the gravimetric heat capacity of the solid phase and ρ_w and C_w are the density and the gravimetric heat capacity of the liquid phase.

Mean weekly air temperatures and temperature amplitudes were calculated from daily air temperatures and from daily minimum and maximum air temperatures for input to the submodel. Simulated temperature profiles gave a good fit for the measured soil temperatures. Simulated temperature profiles are used in other submodels to adjust respiration rates and chemical constants.

MINERAL DISSOLUTION

The rate of dissolution of clay minerals and metal hydrous oxides is surface controlled and is observed to follow zero-order kinetics and can be expressed by the equation (Stumm et al., 1985; Sposito, 1989)

$$d[A] / dt = k$$

where $[A]$ is the aqueous-phase concentration of an ion, and t is time. The parameter k is a function of temperature, pressure, mineral surface area, proton concentration $[H^+]$, and, in the presence of a strong chelator, the ligand concentration $[L^-]$. In a controlled laboratory experiment at constant temperature, pressure, and for a given soil material of some specific mineralogy and mineral surface area, k' is then a function of $[H^+]$ and $[L^-]$

$$k = k'\left[H^+\right]^m\left[L^-\right]^n$$

where m and n are fractional order constants. Combination of the two equations yields

$$d[A]/dt = k'\left[H^+\right]^m\left[L^-\right]^n$$

An experimental set-up was designed to resemble the natural field conditions of mineral dissolution as closely as possible. To avoid the interference of ion exchange with the changes in solution concentrations due to mineral dissolution, the exchange sites of the C horizon samples were first saturated with Ba^{2+}. Hydrochloric acid and 2-hydroxybenzoic acid (salicylic acid) solutions at concentrations of 10^{-3}, 10^{-4}, and 10^{-5} M were used to determine dissolution rates as a function of $[H^+]$ and $[L^-]$. The 2-hydroxybenzoic acid is a six-C ring with a carboxyl and hydroxy group next to each other. The ligand form of this acid (salicylate) forms bidentate (mononuclear) surface complexes (Stumm et al., 1985). The strong chelating effect of salicylate is thought to be representative of the chelating effect of fulvic acids in natural soil solutions of Spodosols. The samples were leached with the acid solutions using a vacuum extractor and time intervals of 1, 6, and 12 h. The samples were leached once again with 0.001 M $BaCl_2$ to remove any adsorbed ions originating from mineral dissolution. The leachates were combined and analyzed for Na^+, K^+, Ca^{2+}, Mg^{2+}, Fe^{2+}, Al^{3+}, and Si^{4+}.

The leachate concentration $[C]$ data were plotted as $[C]$, $\ln[C]$, and $1/[C]$ vs. time to determine the kinetic reaction order. A straight line plot for $[C]$ vs. time indicates zero-order kinetics (Sposito, 1989; Sparks, 1989), whereas a straight line plot for either $\ln[C]$ or $1/[C]$ indicates respectively first- or second-order kinetics. The dissolution kinetics were observed to be of zero order. The k parameters, which are equal to the slope of the straight line plot, were determined for each ion per acid solution treatment. The following dissolution rates (mol/L/h) were obtained by linearly regressing the k parameters with the $[H^+]$ and $[L^-]$ of the different leaching treatments:

$$d\left[Na^+\right]/dt = 2.63E-06 \times \left[H^+\right]^{0.17} \times \left[L^-\right]^{0.03} \qquad r^2 = 0.99$$

$$d\left[K^+\right]/dt = 2.23E-06 \times \left[H^+\right]^{0.11} \times \left[L^-\right]^{0.04} \qquad r^2 = 0.99$$

$$d\left[Ca^{2+}\right]/dt = 8.58E-03 \times \left[H^+\right]^{1.09} \times \left[L^-\right]^{0.03} \qquad r^2 = 0.98$$

$$d\left[Fe^{2+}\right]/dt = 2.55E-03 \times \left[H^+\right]^{1.63} \qquad r^2 = 0.98$$

$$d\left[Al^{3+}\right]/dt = 7.62E-04 \times \left[H^+\right]^{0.80} \times \left[L^-\right]^{0.03} \qquad r^2 = 0.91$$

$$d\left[Si^{4+}\right]/dt = 1.19E-03 \times \left[H^+\right]^{0.87} \times \left[L^-\right]^{0.02} \qquad r^2 = 0.97$$

The Mg^+ leachate concentrations were very low and close to the detection limit and were therefore not used. Examination of the exponents of $[H^+]$ and $[L^-]$ show that the dissolution rates are primarily a function of $[H^+]$ and to a lesser degree a function of $[L^-]$. The dissolution rate of Fe^{2+} was not significantly enhanced by the presence of $[L^-]$.

The submodel for mineral dissolution uses these linear regression equations with their respective k' values to predict yields of individual cations due to dissolution for any given combination of $[H^+]$ and $[L^-]$. These combinations of input variables are derived from the output of the solute movement and chemical speciation submodels. The mineral dissolution submodel is specific for the Typic Haplorthod at our monitoring site. Mineralogy, surface areas, and mineral surface conditions were not specifically defined but were embodied in the k' values of the rate equations.

MICROBIAL DECOMPOSITION OF ORGANIC MATTER

Microbial decomposition is considered to take place in several pools (Bohn et al., 1985; Parton et al., 1987), e.g., structural and metabolic plant remains, active soil organic carbon (SOC) (decomposing plant residues, live microbes), slow SOC (microbial metabolites), passive SOC (humified material), and dissolved organic carbon (DOC). The general equation for the rate of decomposition per pool within a soil environment is

$$dC_x / dt = K_x \times \theta \times T \times C_x$$

where:

C_x = C state variable of Pool x
K_x = decomposition rate for Pool x
θ = volumetric water content
T = soil temperature

The decomposition products are $CO_2 + H_2O$, SOC flowing to other pools, and DOC. The ratio in which these products are produced depends on the volumetric water content.

Dissolved organic carbon is a potentially mobile pool and, hypothetically, is an important factor in the podzolization process. A laboratory experiment was designed to determine rates of microbial CO_2 and DOC production in horizons of the Typic Haplorthod at our field monitoring site. Soil from the Oi, Oa, E, and Bh/Bhs horizons was air dried for 1 d, weighed, and promptly rewet to minimize damage to the microbial population. Per horizon, one series of samples was used to measure the production of CO_2, and another series was used to measure both the CO_2 and DOC production. Samples were rewet or quickly dried (by flushing with dry compressed air) to $\theta = 0.7$, $\theta = 0.5$, $\theta = 0.3$, and $\theta = 0.1$ moisture. The samples used to measure the "CO_2 + DOC" production were prepared by leaching extensively with deionized water using a mechanical vacuum extractor (Concept Engineering, Inc., Lincoln, Nebraska) until no DOC could be detected. During the experiment, CO_2 produced by microbial respiration was flushed out daily for 5 min with CO_2-free compressed air and trapped in a 1.0 M NaOH solution. The CO_2 concentration in these NaOH solutions, trapped as CO_3^{2-}, was measured every 7 d with a Bioscience Inc. (Bethlehem, PA), Model 700, C analyzer. The "CO_2 + DOC" samples also were flushed with deionized water every 7 d to collect the DOC. Blanks were included in all series.

Initially (Day 1–14), the respiration rates were low. Microbial populations then grew exponentially from Day 15 through 35. Following Day 35, the microbial populations reached stable sizes and CO_2 and DOC respiration rates were steady.

The CO_2 respiration rates of the Oi horizon (Day 71–91) were greatest at a volumetric water content of about 0.3 ($m^3 m^{-3}$) and were relatively lower at higher and lower water contents (Fig. 7–4). Reduced oxygen availability and reduced physiological activity may explain the reduced rates for the respectively wetter and drier samples. The CO_2 respiration rates of the Oa, E and Bh/Bhs horizons indicated similar trends, although to a lesser extent. The following polynomials were obtained

$$dCO_{2,Oi} / dt \quad = 38.2 + 89.8\ \theta - 138.8\ \theta^2 \qquad r^2 = 0.99$$

$$dCO_{2,Oa} / dt \quad = 10.0 + 22.0\ \theta - \quad 9.4\ \theta^2 \qquad r^2 = 0.82$$

$$dCO_{2,E} / dt \quad = \ 2.9 + \ 2.1\ \theta - \quad 3.8\ \theta^2 \qquad r^2 = 0.97$$

$$dCO_{2,Bh/Bhs} / dt \quad = \ 2.1 + 10.7\ \theta - \quad 11.9\ \theta^2 \qquad r^2 = 0.50$$

where $dCO_{2,x}/dt$ is the CO_2 respiration rate (μg C d^{-1} g^{-1} soil). The DOC production rates could be described with linear equations (Fig. 7–5)

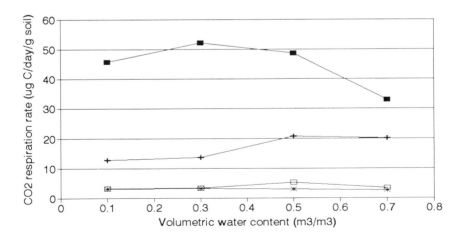

Fig. 7–4.   The CO_2 respiration rates (μg C d^{-1} g^{-1} soil) as a function of volumetric water content ($m^3 m^{-3}$) measured at 20°C.

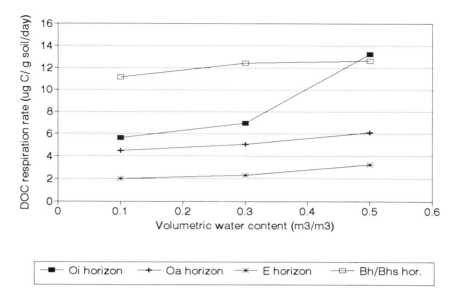

Fig. 7–5. The DOC respiration rates (μg C d^{-1} g^{-1} soil) as a function of volumetric water content (m^{3}m^{-3}) measured at 20°C.

$$d\text{DOC}_{\text{Oi}} / dt = \ \ 2.9 + 19.1\,\theta \qquad\qquad r^2 = 0.88$$
$$d\text{DOC}_{\text{Oa}} / dt = \ \ 4.0 + \ \ 4.1\,\theta \qquad\qquad r^2 = 0.97$$
$$d\text{DOC}_{\text{E}} / dt = \ \ 1.6 + \ \ 3.2\,\theta \qquad\qquad r^2 = 0.92$$
$$d\text{DOC}_{\text{Bh/Bhs}} / dt = 11.0 + \ \ 3.7\,\theta \qquad\qquad r^2 = 0.85$$

where $d\text{DOC}_x/dt$ is the DOC production rate (μg C d^{-1} g^{-1} soil).

These CO_2 and DOC production rate equations were established with respiration data measured at 20°C. For use in the microbial decomposition submodel, these rates need to be adjusted for the particular soil temperature of each simulated layer. A Q_{10}-type temperature response was assumed for the microbial activity, with Q_{10}, the factor by which the rate constant changes over a 10°C interval, equal to two (Johnsson et al., 1987; Hutson & Wagenet, 1992). The temperature correction factor, T_{corr}, was calculated as

$$T_{\text{corr.}} = Q_{10}{}^{0.1\left(T\text{soil} - T\text{base}\right)}$$

where Q_{10} was assumed to be two, T_{soil} is the soil temperature (°C), and T_{base} = 20°C. The predicted CO_2 and DOC yields for a particular layer with a soil temperature of 10°C will be reduced by 50%. The soil temperature and volumetric water content data used as input to the microbial decomposition submodel were derived from the output of the soil temperature and water movement submodels. Output from the microbial decomposition submodel was used as ligand concentration input for the mineral dissolution and solute movement submodels.

ION EXCHANGE

Ion exchange was recognized as an important process controlling the chemistry of the soil solution. A submodel was developed to represent this process in our pedo-dynamic model. Many different exchange equations have been proposed to describe the exchange between cations of unequal charge. The most general form of an exchange equation is based on the mass action equation (Bohn et al., 1985), e.g.,

$$3CaX + 2Al^{3+} = 2AlX + 3Ca^{2+}$$

with the selectivity coefficient

$$K = \frac{\left(AlX\right)^2 \left(Ca^{2+}\right)^3}{\left(CaX\right)^3 \left(Al^{3+}\right)^2}$$

where X denotes the ion in the adsorbed phase. It is assumed that the ions in the adsorbed phase behave as if being in solution. Vanselow (1932) substituted the mole fractions of the exchangeable ions in the above exchange equation. The mole fraction, e.g., $N(Al)$, is defined as

$$N\left(Al\right) = n_{Al} / \left(n_{Al} + n_{Ca}\right)$$

where n is mole of exchangeable ions per grams of soil. The Gapon equation is widely used for Na^+–Ca^{2+} exchange (Bohn et al., 1985). The mass action equation for Ca^{2+}–Al^{3+} exchange, which is an important exchange reaction in a Spodosol, is written as

$$CaX + 2/3Al^{3+} = \left(Al\right)_{2/3} X + Ca^{2+}$$

with the exchange coefficient

$$K_{Gapon} = \frac{\left[\left(Al\right)_{2/3} X\right]\left[Ca^{2+}\right]}{\left[CaX\right]\left[Al^{3+}\right]^{2/3}}$$

Exchange equations based on mass action equations assume that the activities of the adsorbed ions are proportional to their equivalent or mole fraction in the adsorbed phase. However, data from a Ca^{2+}–Al^{3+} exchange experiment on Montmorillonite at low pH showed a large preference of the clay for the trivalent ion (McBride & Bloom, 1977). An exchange model based on a statistical thermodynamic approach was derived to better describe the activity of Al^{3+} at the mineral surface. The equation relates the Al^{3+} in solution and the equivalent fraction of adsorbed Al^{3+} as

$$\left(Al^{3+}\right) = K_{McBride} N\left(Al\right) / \left[1 - N\left(Al\right)\right]$$

This equation emphasizes the lack of dependence of Al^{3+} adsorption on the $[Ca^{2+}]$ in solution, a result of the more solutionlike nature of adsorbed Ca^{2+} in comparison to the strongly adsorbed Al^{3+}.

To represent a field situation, soil samples from the major horizons were equilibrated by slowly leaching with different base ion and $AlCl_3$ solutions. Due to the concentration charge effect (a preference of adsorption of tri- and divalent ions as compared to monovalent ions at low electrolyte concentrations), which applies to the study site (excessive leaching, low electrolyte concentrations), Al^{3+}, and to a lesser degree Ca^{2+} and Mg^{2+}, will dominate the exchange sites (McBride, 1994). Also, from a pedogenetic point of view, it is of interest to simplify the system to an exchange between Al^{3+} and the summation of the base cations. Table 7–1 shows the average ratios of base cations per horizon as determined by analysis of soil solutions sampled every 4 wk at our monitoring field site from April through November 1991. Samples of each horizon were saturated with the corresponding base ratios with the use of a vacuum extractor. All exchange sites were assumed to be occupied by base cations. Equilibrium solutions were made by mixing 0.001 M base ratio solutions with 0.001 M $AlCl_3$ solution. The samples were equilibrated by leaching three times with 50 mL of 0.001 M equilibrium solutions in 12-h runs. Ethanol (95%) was used to wash out excess equilibrium solution. Next, the Al^{3+} and base cations on the exchange sites were exchanged by Ba^{2+} by leaching two times with 0.10 M $BaCl_2$ solution in 12-h runs. These leachates were analyzed for Al^{3+} and base cations with an inductively coupled plasma optical emission spectrometry (ICP).

The ion exchange isotherms of the Oi and Bh horizons are presented in Fig. 7–6 and 7–7. The shape of the isotherms depends in part on the "history" of the system, i.e., the exchange sites were first saturated with base cations. Initially most parent materials have a relatively high base status which over time in a leaching environment generally decreases by soil forming processes. The recovery of Al^{3+} from samples equilibrated with Al^{3+} mole fractions of 0.8 and 1.0 is relatively low, probably due to the failure of Ba^{2+} to exchange all Al^{3+}. Due to the low organic matter and clay percentages, the cation exchange capacity (CEC) of the E, Bs, 2BC, and 3C horizons were too low to obtain detectable exchange data. The exchange coefficients, K, for the three exchange equations are presented in Table 7–2. The exchange coefficients based on the mass action equations, $K_{Vanselow}$ and K_{Gapon}, remain relatively constant for the organic Oi and Oa horizons as compared to $K_{McBride}$. However, for the mineral Bh and Bhs horizons, the exchange coefficients of the McBride equation are essentially constant as compared to $K_{Vanselow}$ and K_{Gapon}.

The CEC of the O horizons is primarily the result of the ionization of functional groups on organic matter. Exchange reactions on organic matter can be

Table 7–1.   Average base cation ratios [e.g., $Ca^{2+}/(K^++Na^++Ca^{2+}+Mg^{2+})$] of soil solutions from April through November 1991.

Horizon	K^+	Na^+	Ca^{2+}	Mg^{2+}	Total
Oa	0.19	0.12	0.59	0.10	1.00
E	0.12	0.29	0.49	0.09	1.00
Bhs	0.13	0.31	0.45	0.11	1.00
Bs	0.31	0.26	0.34	0.10	1.00
3C	0.07	0.43	0.41	0.10	1.00

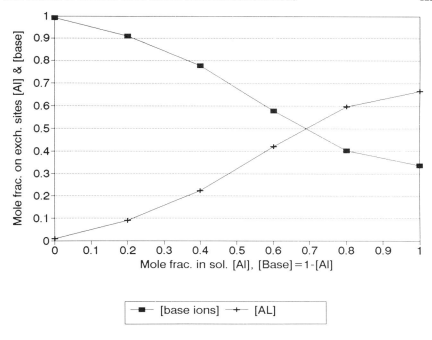

Fig. 7–6.   Ion exchange isotherm for "base ion mix" → Al^{3+} exchange on Oi horizon soil material.

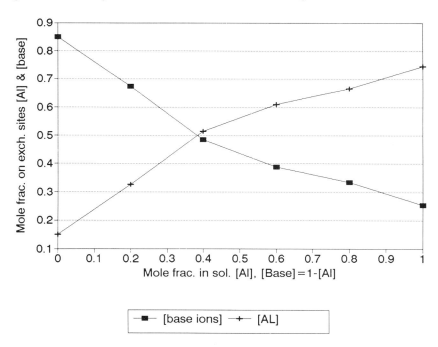

Fig. 7–7.   Ion exchange isotherm for "base ion mix" → Al^{3+} exchange on Bh horizon soil material.

Table 7–2.   Exchange coefficients of the Vanselow, Gapon, and McBride exchange equations for
Al^{3+} adsorption on base ion saturated samples from different horizons.

Horizon	N(Al)	$K_{Vanselow}$	K_{Gapon}	$K_{McBride}$
Oi	0.089	0.13	0.15	2.05
	0.223	0.14	0.21	1.40
	0.423	0.17	0.27	0.82
	0.596	0.07	0.23	0.54
	0.664	0	0	0.51
Oa	0.113	0.23	0.20	1.58
	0.263	0.23	0.26	1.12
	0.416	0.15	0.27	0.84
	0.577	0.06	0.21	0.59
	0.730	0	0	0.37
Bh	0.326	4.46	0.76	0.41
	0.515	3.14	0.78	0.38
	0.611	1.13	0.59	0.38
	0.665	0.15	0.31	0.40
	0.745	0	0	0.34
Bhs	0.404	9.86	1.06	0.30
	0.584	6.41	1.04	0.28
	0.660	1.96	0.73	0.31
	0.743	0.73	0.55	0.22
	0.782	0	0	0.35

Table 7–3.   Exchange coefficients used in the ion exchange submodel.

Horizon	Exch coefficient	Mean	SD
Oi	$K_{Vanselow}$	0.10	0.07
Oa	$K_{Vanselow}$	0.13	0.10
Bh	$K_{McBride}$	0.38	0.03
Bhs	$K_{McBride}$	0.29	0.05

described by a mass action type equation using $K_{Vanselow}$ or K_{Gapon}. The B horizons
have a higher clay content and are rich in sesquioxides. These horizons are therefore modeled with the McBride equation. Exchange coefficients used in the submodel for horizons with significant exchange properties are presented in Table
7–3. Given solution chemistry and solute movement fluxes as input, the ion
exchange submodel calculated the change in solution chemistry for each layer on
a daily basis.

CHEMICAL SPECIATION AND PRECIPITATION

Since the objective of the pedodynamic model is to simulate the chemistry
and movement of the major chemical components, a chemical speciation and precipitation submodel was needed. The chemical equilibrium program MINEQL[+]
(Westall et al., 1976; Schecher & McAvoy, 1991) was selected to calculate chemical speciation and the precipitation of solid phases.

DISCUSSION

The submodels presented in this paper were used as building blocks in the pedodynamic ORTHOD model. The successful application of this model to the simulation of C dynamics in three Typic Haplorthod soil profiles, has been presented elsewhere (Hoosbeek & Bryant, 1994). As an example, the measured and simulated DOC concentrations of the E and Bhs horizons are presented in Fig. 7–8. The CO_2 and DOC fluxes were simulated daily. Simulated net DOC fluxes for each layer over a 365-d period show large net DOC fluxes into the Bh horizon and relatively small fluxes into the Bs and BC horizons (Fig. 7–9).

Using the investigated Typic Haplorthod several podzolization theories were tested with the ORTHOD model, and presented elsewhere (Hoosbeek & Bryant, 1994, unpublished data). The model successfully simulated the chemistry and movement of Al, Si, DOC, and other chemical species involved in the podzolization process. The net Al balance over a 365-d period per layer showed significant losses of Al from the E horizon, losses from the Bh and Bhs, and accumulations of Al in the Bs and BC horizons (Fig. 7–10).

The present paradigm of pedology is based on the *Factors of Soil Formation* (Jenny, 1941). However, some pedological or environmentally related questions cannot be answered by applying the factors of soil formation. How much DOC— produced in the O horizon—will be sequestered in the B horizons, and how much DOC will leave the profile per year? Which mechanisms regulate Al activity in the root zone? Which podzolization mechanisms govern the accumulation of Al in the Bs horizon and how do these rates vary throughout the year? The pedodynamic approach allows for the utilization of current knowledge in soil science through the use of interchangeable submodels. A pedodynamic model can be gradually improved as individual submodels are refined. For instance, a water

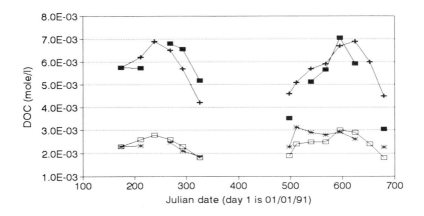

Fig. 7–8.   Measured and simulated DOC concentrations of the E and Bhs horizons.

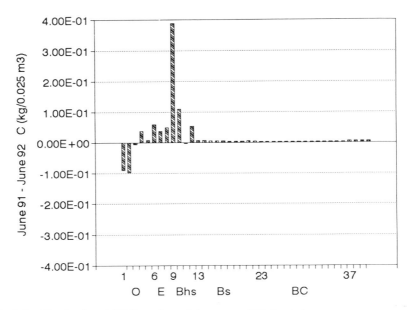

Fig. 7–9.   Simulated net DOC fluxes for each layer over a 365-d period.

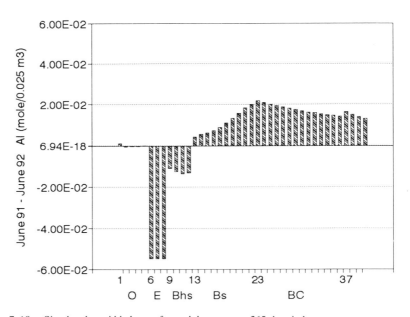

Fig. 7–10.   Simulated net Al balances for each layer over a 365-d period.

movement submodel based on the Richards equation might be replaced by a model capable of dealing with preferential flow without affecting the other submodels of a pedodynamic model. The choice of submodels also depends on the research objectives.

The ORTHOD model quantitatively described the dynamics of a representative pedon of a Haplorthod at the horizon (i–1) and molecular (i–3) levels. In its present stage, the ORTHOD model is limited to a one-dimensional, vertical, representation of soil processes in a pedon. A pseudo multidimensional representation may be obtained by applying the model to adjacent points in a soil landscape. A truly two- or three-dimensional pedodynamic approach would have to include multidimensional soil process algorithms, e.g., vertical and lateral solute movement.

Although the *Factors of Soil Formation* and the pedodynamic models differ in approach (qualitative vs. quantitative and functional vs. mechanistic), they complement each other in respect to organizational hierarchy (Fig. 7–1). Depending on the level of the soil system under investigation, the models can be used complementary in pedological research.

REFERENCES

Bohn, H.L., B.L. McNeal, and G.A. O'Connor. 1985. Soil chemistry. Wiley-Interscience, New York.

Campbell, G.S. 1985. Soil physics with BASIC; transport models for soil-plant systems. Elsevier, Amsterdam.

Hoosbeek, M.R., and R.B. Bryant. 1992. Towards the quantitative modeling of pedogenesis—A review. Geoderma 55:183–210.

Hoosbeek, M.R., and R.B. Bryant. 1994. Modeling the dynamics of organic carbon in a Typic Haplorthod. *In* R. Lal et al. (ed.) Soils and global change. Adv. Soil Sci., Lewis Publ., Chelsea, MI. (In Press.)

Hutson, J.L., and A. Cass. 1987. A retentivity function for use in soil-water simulation models. J. Soil Sci. 38:105–113.

Hutson, J.L., and R.J. Wagenet. 1992. LEACHM; leaching estimation and chemistry model. Res. Ser. no. 92-3. Dep. Soil, Crop and Atmos. Sci., Cornell Univ., Ithaca, NY.

Jenny, H. 1941. Factors of soil formation—A system of quantitative pedology. McGraw-Hill, New York.

Johnsson, H., L. Bergstrom, P.E. Jansson, and K. Paustian. 1987. Simulated nitrogen dynamics and losses in a layered agricultural soil. Agric. Ecosyst. Environ. 18:333–356.

McBride, M.B. 1994. Environmental chemistry of soils. Oxford Univ. Press, New York.

McBride, M.B., and P.R. Bloom. 1977. Adsorption of aluminum by a smectite: II. An Al^{3+}–Ca^{2+} exchange model. Soil Sci. Soc. Am. J. 41:1073–1077.

Parton, W.J., D.S. Schimel, C.V. Cole, and D.S. Ojima. 1987. Analysis of factors controlling soil organic matter levels in Great Plain Grasslands. Soil Sci. Soc. of Am. J. 51:1173–1179.

Schecher, W.D., and D.C. McAvoy. 1991. MINEQL+: A chemical equilibrium program for personal computers. User's manual. Version 2.1. Environ. Res. Software, Edgewater, MD.

Simonson, R.W. 1959. Outline of a generalized theory of soil genesis. Soil Sci. Soc. Am. Proc. 23:152–156.

Sparks, D.L. 1989. Kinetics of soil chemical processes. Acad. Press, San Diego, CA.

Sposito, G. 1989. The chemistry of soils. Oxford Univ. Press, New York.

Stumm, W., G. Furrer, E. Wieland, and B. Zinder. 1985. The effects of complex-forming ligands on the dissolution of oxides and aluminosilicates. p. 55–74. *In* J.I. Drever (ed.) The chemistry of weathering. D. Reidel Publ. Co., Dordrecht, the Netherlands.

Tillotson, W.R., C.W. Robbins, R.J. Wagenet, and R.J. Hanks. 1980. Soil water, solute, and plant growth simulation. Bull. 502. Utah State Agric. Exp. Stn.

van Genuchten, M. Th., and P.J. Wierenga. 1986. Solute dispersion coefficients and retardation factors. p. 1025–1054. *In* A. Klute (ed.) Methods of soil analysis. Part 1. Agron. Monogr. 9. ASA and SSSA, Madison, WI.

Vanselow, A.P. 1932. Equilibria of the base-exchange reactions of bentonites, permutites, soil colloids, and zeolites. Soil Sci. 33:95–113.

Wagenet, R.J. 1986. Water and solute flux. p. 1055–1088. *In* A. Klute (ed.) Methods of Soil Analysis. Part 1. Agron. Monogr. 9. ASA and SSSA, Madison, WI.

Wagenet, R.J., and J.L. Hutson. 1987. LEACHM: Leaching estimation and chemistry model. Continuum Vol. 2. Water Resour. Inst., Cornell Univ., Ithaca, NY.

Westall, J.C., J.L. Zachary, and F.M.M. Morel. 1976. MINEQL, A computer program for the calculation of chemical equilibrium composition of aqueous systems. Tech Note18, Dep. Civil Eng., Massachusetts. Inst. Technol. Cambridge, MA.

8

Quantitative Modeling of Soil Forming Processes in Deserts: The CALDEP and CALGYP Models

G.M. Marion

U.S. Army Corps of Engineers
Hanover, New Hampshire

William H. Schlesinger

Duke University
Durham, North Carolina

Many models have been developed to simulate $CaCO_3$ (calcite) formation in soils. Mechanistic models that focus primarily on the long-term formation of $CaCO_3$ are summarized in Table 8–1. These models differ with respect to: (i) the mechanism of $CaCO_3$ formation, (ii) Ca sources, (iii) the inclusion of chemical kinetics, (iv) spatial and temporal variability in pH and CO_2, (v) the type of rainfall model, and (vi) the inclusion of other minerals in the model. Calcium carbonate may precipitate in soil as the result of downward movement of Ca with percolating water from rainfall (the "illuvial" process); this is clearly the dominant process in the desert Southwest (Gile et al., 1981; Birkeland, 1984; Machette, 1985; Hendricks, 1991; McFadden et al., 1991). Alternatively, $CaCO_3$ may precipitate in the soil profile as the result of upward movement of Ca with evaporating waters from shallow water tables (the "evaporative" process); this process is locally important in a variety of habitats (Amundson & Lund, 1987; Knuteson et al., 1989; Marion et al., 1991). Calcium sources for the formation of $CaCO_3$ include weathering of both carbonate and noncarbonate minerals, groundwater, rainfall, and dust. The dominant Ca sources in southwestern deserts are dust and rainwater for soils developing on noncalcareous parent materials (Gile et al., 1981; Birkeland, 1984; Machette, 1985; Harden et al., 1991; McFadden et al., 1991). Only the Rogers and McFadden models consider chemical kinetics as a possible rate-limiting step in the dissolution of calcite, all other models assume that the solution phase is in chemical thermodynamic equilibria at all times. Most evidence suggest that calcite kinetics are fast and chemical thermodynamic equilibrium for calcite is a reasonable assumption for desert soils (McFadden et al., 1991). Most models consider spatial variability of CO_2 and pH, but few consider

Table 8–1. A summary of features in published CaCO$_3$ simulation models.

Reference	Formation mechanism[†]	Ca source[‡]	Chemical kinetics[§]	pH and CO$_2$ variability		Rainfall		Other minerals[§]
				Spatial[§]	Temporal[§]	Model[¶]	Time Step[#]	
Arkley (1963)	I	M	N	N	N	D	M	N
Rogers (1980)	I	M	Y	Y	N	D	D	Y
Zelichenko & Sokolenko (1982)	E	G	N	Y	N	—	—	N
McFadden (1982)								
McFadden & Tinsley (1985)	I	P, D	Y	Y	N	D	M	N
Marion et al. (1985)	I	P, D	N	Y	Y	S	D	N
Mayer et al. (1988)	I	D	N	Y	N	D	M	N
Marion & Schlesinger (this chapter)	I	P, D	N	Y	Y	S	D	Y

[†] I = illuvial, E = evaporative.

[‡] M = mineral weathering, G = groundwater, P = precipitation, D = dust.

[§] N = no, Y = yes.

[¶] D = deterministic, S = stochastic.

[#] M = monthly, D = daily.

temporal variability (Table 8–1). Most rainfall models are deterministic (using the same average rainfall pattern year-after-year) with monthly time steps. Only Rogers (1980) considers the precipitation-dissolution of a broad suite of minerals in addition to calcite. The CALGYP model (this chapter) considers both calcite and gypsum ($CaSO_4 \cdot 2H_2O$) formation.

Several simulation models have been developed to track short-term salt movement through soils that include the precipitation and dissolution of calcite and gypsum. Examples include the Dutt et al. (1972) model, the SALTFLOW model (Robbins et al., 1980; Dudley et al., 1981), and the Simunek-Suarez model (Simunek & Suarez, 1992). These models typically have sophisticated routines for both the physics of water movement and chemical processes. To date, however, these models have not been used for simulating long-term pedogenesis because they are computationally slow.

The CALDEP model was developed to predict calcite formation in desert soils (Marion et al., 1985). Recently, we modified this model to consider the formation of both calcite and gypsum (the CALGYP model). The objective of this paper is to demonstrate, with the CALDEP and CALGYP models, how simulation models may be usefully applied to studies of soil chemistry and pedogenesis. Applications of the CALDEP-CALGYP models to: (i) a fundamental question in soil chemistry, the pH-sulfate relationship, (ii) the control of calcite and gypsum formation in desert soils, (iii) paleo-processes, and (iv) spatial and temporal variability, will be examined.

MODEL STRUCTURE

The CALDEP and CALGYP models share the same basic structure except that CALGYP includes additional routines to describe sulfate chemistry. Because the CALDEP model has been published (Marion et al., 1985), only a brief overview of the model structure is given in this paper. Detailed descriptions are given for new sulfate routines introduced into CALDEP to produce CAL-GYP. The CALGYP(CALDEP) model has five components: soil parameterization, chemical thermodynamic relations, a stochastic precipitation model, an evapotranspiration model, and subroutines to handle water, Ca, and sulfate flux through the soil.

The CALGYP is a compartment model that can be parameterized to contain 1 to 10 layers (Fig. 8–1). Model inputs for each layer include layer thickness, bulk density, water contents initially and at 0.01 MPa (field capacity) and 1.5 MPa (permanent wilting point), initial soil calcite and gypsum contents, initial concentrations of soluble Ca and SO_4-S, and initial soil pH. A model for soil CO_2, developed from a study near Tucson, Arizona (Parada et al., 1983), allows CO_2 concentration to vary spatially and seasonally (Marion et al., 1985). Atmospheric inputs include the Ca and SO_4-S content of rainfall and dust.

Chemical thermodynamic relations include the equilibrium constants and their temperature dependence for: CO_2 distribution between the soil gas and aqueous phases (Henry's Law constant), the first and second dissociation constants for carbonic acid, the stability constant for the $CaSO_4$ ionpair, and the solubility products for soil calcite and gypsum.

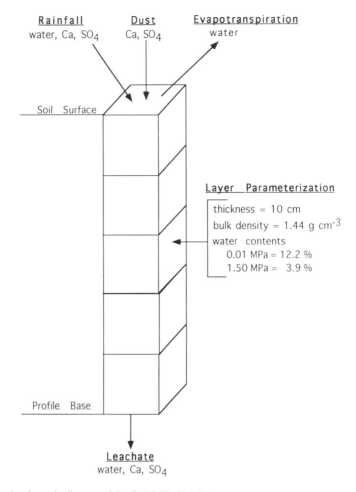

Fig. 8–1. A schematic diagram of the CALDEP-CALGYP models.

Critical to the performance of these models is the mechanism for estimating soil pH, because of the strong influence of pH on $CaCO_3$ solubility. For a pure $CaCO_3$-$CaSO_4$ solution in the pH range of 7 to 8.5, the following charge balance holds

$$2\left[Ca^{2+}\right] = \left[HCO_3^-\right] + 2\left[CO_3^{2-}\right] + 2\left[SO_4^{2-}\right] \tag{1}$$

where brackets refer to concentrations. For a system in equilibrium with solid phase $CaCO_3$, this equation can be rewritten as

$$\frac{2K_{cal}\left(H^+\right)^2}{\gamma_{Ca}K_2K_1K_HP_{CO_2}} = \frac{K_HK_1P_{CO_2}}{\left(H^+\right)\gamma_{HCO_3}} + \frac{2K_HK_1K_2P_{CO_2}}{\left(H^+\right)^2\gamma_{CO_3}} + 2\left[SO_4^{2-}\right] \tag{2}$$

where K_{cal} is the $CaCO_3$ solubility product, K_H is the Henry's Law constant, K_1 and K_2 are the first and second dissociation constants for carbonic acid, (H^+) is the hydrogen ion activity, P_{CO2} is the partial pressure of CO_2, and the γ's are activity coefficients. The activity coefficients are estimated from ionic strength by successive approximations using the Davies equation (Davies, 1962). Given the P_{CO2} and the SO_4-S concentration, Eq. [2] is solved for (H^+), which is then used to control $CaCO_3$ solubility.

For a system in equilibrium with both solid phase calcite and gypsum, the sulfate term in brackets (Eq. [1] and [2]) may be replaced by

$$\frac{K_{gyp} K_H K_1 K_2 P_{CO_2}}{K_{cal} \gamma_{SO_4} \left(H^+\right)^2} \qquad [3]$$

where K_{gyp} is the solubility product of gypsum. Compared to equilibrium with pure calcite, simultaneous equilibrium of gypsum and calcite significantly depresses the equilibrium pH (Fig. 8–2A) and increases the equilibrium Ca concentrations (Fig. 8–2B). In the CALGYP model, Eq. [2] is used to estimate solution pH (after $CaCO_3$ begins to precipitate), so that soil pH is a continuous function of soil sulfate concentration. Empirical justification for this assumption is discussed in the Applications section.

The stochastic rainfall model controls input of water and is based on probability distributions for interarrival time (the number of days between rainfall events) and the rainfall amounts for specific seasons at specific sites. Sites currently in the model include: Yuma, Phoenix, and Tucson in Arizona; Albuquerque, Roswell, and Clayton in New Mexico; and El Paso, Texas. A random number generator is used to select the interarrival times and rainfall amounts for each year from the cumulative probability distributions. This stochastic model is designed to reproduce the long-term average annual rainfall and the variability in annual rainfall for a specific site (Marion et al., 1985).

The evapotranspiration model controls the loss of water and consists of three steps. First, potential evapotranspiration is calculated using Thornthwaite's equation. Second, Thornthwaite's potential evapotranspiration is converted to pan evaporation using a derived, empirical relationship for southwestern deserts. And third, actual evapotranspiration is calculated as a function of soil moisture and pan evaporation (Marion et al., 1985). Calibration of the third step is based on field measurements from a creosote bush *(Larrea tridentata)* site at the Jornada Desert Long-Term Ecological Research site near Las Cruces, New Mexico.

A daily time-step is used to assess the flux of water and solutes through the soil. All rainfall is assumed to enter the uppermost soil layer (Fig. 8–1), the model ignores vegetative interception of rainfall and surface runoff. Only saturated flow of water is considered in these models. If the water-holding capacity of a layer is exceeded, excess water moves into progressively deeper layers. Water flux beyond the base of the soil profile is treated as leachate and is assumed lost from the system. Solutes are assumed to move with the mass flow of water. Water that enters a given layer is mixed with the pre-existing water and solutes are chemically equilibrated with the solid and gas phases. Therefore, the excess water that passes through a given layer contains an equilibrated concentration of solutes

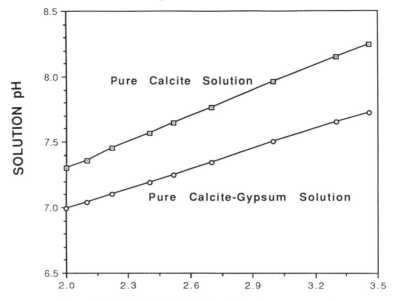

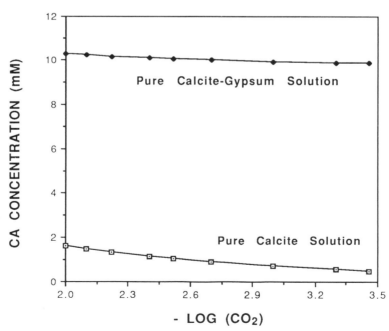

Fig. 8–2.   The theoretical (A) pH and (B) Ca concentrations for pure calcite and calcite-gypsum solutions as functions of the partial pressure of CO_2 at 25°C.

before passing to the deeper layers. During drying cycles, water is first extracted from the surface layer and then from progressively deeper layers using the previously mentioned evapotranspiration model.

APPLICATIONS

The pH-Sulfate Relationship

The only major "untested" new concept in the CALGYP model is the equation for estimating pH (Eq. [2]). Although it is easy to demonstrate that the presence or absence of gypsum in theoretical pure solutions affects pH (Fig. 8–2A), it remains to be shown that the presence of soluble $[SO_4^{2-}]$ plays a role in controlling the pH of field soils, which are inherently more complex than pure solutions.

To evaluate the pH-sulfate relationship, data were selected from two field studies. The first study was in the eastern Mojave Desert in southern California, this soil was classified as a Calciorthid (Schlesinger, 1985). The second study was along the Tanana River of interior Alaska, this soil was classified as a Typic Cryofluent (Marion et al., 1993). Both soils contain calcite. Gypsum is common throughout the profile in the Alaska soil but is present only infrequently in the Mojave soil.

Using measured K_{cal}, P_{CO2}, and $[SO_4^{2-}]$ concentrations for each soil in Eq. [2] led to a good fit to the experimental data for the Mojave soil but an underestimation of pH for the Alaska soil (Fig. 8–3, dashed lines). Apparently Eq. [2] does not adequately describe the charge balance, especially for the Alaska soil. For both sites, there was an excess of cations (Mg, K, Na) over anions (Cl) not explicitly considered in the model (Eq. [1]). Adding the average difference in these ions ("excess cations") as a constant to the left side of Eq. [2] before solving for (H^+) led to a reasonable fit to the experimental data for both sites (Fig. 8–3, solid lines).

Evidence for a pH-sulfate relationship in soils is sparse. Schlesinger (1985) noted that soil pH was positively correlated to $p[SO_4^{2-}]$ in the Mojave Desert study, pH decreased significantly as gypsum solubility was approached (Fig. 2 in Schlesinger, 1985). Reheis (1987) in a pedogenetic study of gypsic soils in Wyoming found that soil pH increased with increasing soil age up to 30,000 yr, during this period $CaCO_3$ was the dominant secondary mineral. However, between 30,000 and 60,000 yr, when gypsum became the dominant secondary mineral, soil pH gradually declined (Fig. 10 in Reheis, 1987). Both studies support the argument advanced in CALGYP (Eq. [2], Fig. 8–3) that the presence of sulfate controls, in part, soil pH.

In the following simulations, we use Eq. [2] without modification which is equivalent to assuming a pure $CaCO_3$–$CaSO_4$ system. Use of the CALGYP model for specific sites where the charge balance may differ significantly from that assumed in Eq. [2] can apparently be accommodated by adding a constant term to either the left (excess cations) or right (excess anions) of Eq. [2].

This specific case illustrates how modeling encourages integration of empirical and theoretical knowledge which frequently identifies gaps in our knowledge (the pH-sulfate relationship). The ultimate outcome of this modeling exercise is a better integrative understanding of processes controlling soil chemistry and pedogenesis.

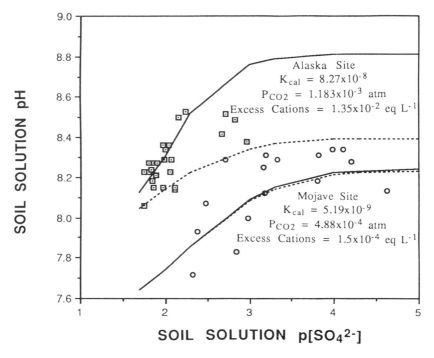

Fig. 8–3. The relationship between soil solution pH and $p[SO_4^{2-}]$ for an Alaska soil (Marion et al., 1993) and a Mojave soil (Schlesinger, 1985). The dashed lines are the theoretical lines using Eq. [2]. The solid lines are the theoretical lines after adding "excess cations" as a constant to the left of Eq. [2].

Calcite-Gypsum Formation

We ran simulations to evaluate the role of sulfate on calcite and gypsum precipitation in desert soils. Soil properties (e.g., bulk density, water holding capacity) were parameterized identically to those used in the original CALDEP paper (Fig. 8–1; Marion et al., 1985). We used climatic drivers (a stochastic rainfall model and mean monthly temperatures) for Tucson, Arizona (mean annual rainfall = 28.4 cm, mean annual temperature = 19.7°C). The P_{CO_2} data were from the Parada et al. (1983) study near Tucson. We used a Ca input rate of 0.86 g Ca m^{-2} yr^{-1} and a SO_4-S input rate of 0.08 g S m^{-2} yr^{-1} based on measured inputs from a Wyoming site (Reheis, 1987). If all of this Ca precipitates as $CaCO_3$, this is equivalent to an annual $CaCO_3$ formation rate of 2.15 g $CaCO_3$ m^{-2} yr^{-1} which is in the midrange for sites in the desert Southwest (Machette, 1985; Schlesinger, 1985; Marion, 1989).

Simulations were run both with and without sulfate additions. In both cases, rainfall was sufficient to prevent formation of calcite in the surface 20 cm (Fig. 8–4). Gypsum precipitated more deeply in the soil profile than calcite, as one would expect given its greater solubility (Lindsay, 1979) and from field observations (Birkeland, 1984; Schlesinger, 1985; Reheis, 1987; Harden et al., 1991;

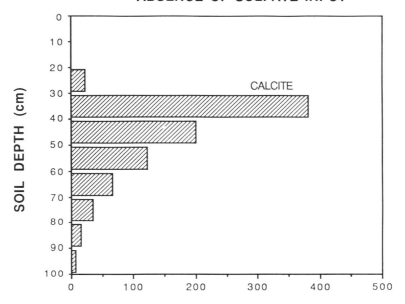

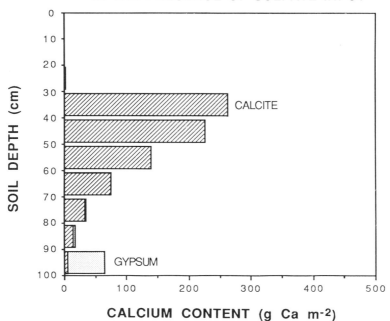

Fig. 8–4.   Simulations (1000 yr) of calcite and gypsum formation in the (*A*) absence of sulfate input and (*B*) presence of sulfate input using a stochastic rainfall model.

McFadden et al., 1991). For a 1000-yr simulation in the absence of sulfate, 98.8% of the added Ca precipitated as calcite, 0.8% leached past the base of the profile, and 0.4% accumulated as soluble Ca (Fig. 8–4A). In the presence of sulfate, 87.8% of the Ca precipitated as calcite, 7.6% precipitated as gypsum, 3.0% leached past the base of the profile, and 1.6% accumulated as soluble Ca (Fig. 8–4B). The greater solubility of Ca in the presence of sulfate (1.6 vs. 0.4%) is an indirect effect of sulfate on decreasing soil pH (Fig. 8–2A, 8–3) which increases Ca solubility (Fig. 8–2B).

The increased Ca solubility also is reflected in the greater mobility of Ca in the presence of sulfate. In the presence of sulfate, 57.0% of the added Ca precipitated as calcite in the 20- to 50-cm layers (Fig. 8–4B). In the absence of sulfate, 70.1% of the added Ca precipitated as calcite in the 20- to 50-cm layers (Fig. 8–4A). A greater proportion of the total Ca leached past the base of the 1.0-m soil profile in the presence of sulfate (3.0%) than in the absence of sulfate (0.8%).

Of the total sulfate added to the soil in the 1000-yr simulation, 65.7% precipitated as gypsum, mostly in the 90- to 100-cm layer (Fig. 8–4B), 23.7% leached past the base of the soil profile, and 10.6% remained in the soil profile as soluble sulfate. Of the total water added to the soil in this 1000-yr simulation (29,006 cm), only 5.0 cm (0.017%) leached past the base of the soil profile; but this small amount of leachate carried a significant amount of Ca (3.0%) and sulfate (23.7%). The latter leachate (output) percentages translate into a molar Ca/SO$_4$ ratio of 1.09. This ratio is in marked contrast to the molar input ratio of 8.60. Over time, Ca is preferentially removed from solution by precipitation as CaCO$_3$ and SO$_4$ is progressively concentrated in the deeper layers until gypsum precipitates (Fig. 8–4). What passes as leachate beyond 1.0 m is predominantly a CaSO$_4$ solution.

Our theoretical results support the argument that sulfate in the soil solution increases the mobility of Ca and CaCO$_3$ in the soil (cf., Fig. 8–4A, 4B). Reheis (1987), on the other hand, argued that the mobility of CaCO$_3$ should be restricted in gypsic soils because of the "common ion effect"; her calculations, however, did not consider the critical effect of pH in controlling the solubility of CaCO$_3$. Our theoretical calculations (Fig. 8–2) and laboratory data (Fig. 8–3) indicate that the effect of sulfate on pH and CaCO$_3$ solubility is sufficient to increase CaCO$_3$ mobility in the presence of sulfate (Fig. 8–4). Calcium carbonate formation in soils is an inherently complex process with many strong interactions especially between climatic and edaphic factors. The ability to handle complexity is one of the many positive attributes of simulation models.

Paleo-Processes

Simulation models are used to evaluate the role of past changes in state factors, such as climate and vegetation, on soil processes. Field evidence strongly supports polygenetic profile development (a significant difference between the present and the past) for soils of the desert Southwest. For example, many features of desert soil profiles, such as CaCO$_3$ and clay horizons, probably formed under earlier "wetter" climates (Gile et al., 1981; Machette, 1985; McFadden & Tinsley, 1985; Harden et al., 1991; McFadden et al., 1991). During the late Pleistocene (20 000 YBP) in the Southwest, woodlands existed over much of the contemporary desert. A major climatic change has been implied to explain changes in vegetation during the Holocene (Wells, 1966, 1979; Van Devender & Spaulding, 1979; Galloway, 1983).

In the CALDEP paper (Marion et al., 1985), we evaluated current climate and three hypothetical Pleistocene climates as model drivers. The first Pleistocene scenario was the cold-dry hypothesis with temperatures 10°C colder than present and rainfall 80% of present rainfall (Galloway, 1983). The second Pleistocene scenario was the cool-wet (summer) hypothesis with temperatures 5°C colder and rainfall 67% greater than present, with the increased rainfall coming during the summer months (Wells, 1966, 1979). The third Pleistocene scenario was the cool-wet (winter) hypothesis, which was similar to the second hypothesis except that increased rainfall came in the winter (Van Devender & Spaulding, 1979).

Current climatic drivers predicted a shallower depth of calcite deposition than is found in most southwestern soils (Fig. 8–5). Only the cool, wet (winter) Pleistocene hypothesis could account for polygenetic profile development and the depth of the dominant $CaCO_3$ horizon which is known to be Pleistocene in age (Marion et al., 1985). McFadden and Tinsley (1985) also concluded that a wet

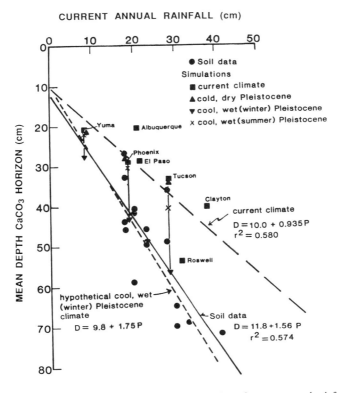

Fig. 8–5. The mean depth of the $CaCO_3$ horizon as a function of current annual rainfall (adapted with permission from Marion et al., 1985).

pluvial climate would not be conducive to polygenetic calcic soil development if increased rainfall came in the summer. However, they concluded that a polygenetic profile could develop as a result of change from a cold Pleistocene climate to a much warmer ($+8°C$) Holocene climate with no associated change in rainfall. Their cold hypothesis is similar to our cold-dry Pleistocene hypothesis (Fig. 8–5), except that we assumed Pleistocene rainfall was 80% of present rainfall, and McFadden and Tinsley (1985) assumed Pleistocene rainfall was the same as present rainfall. It is clear that our cold-dry Pleistocene hypothesis with a reduced rainfall is not conducive to polygenetic soil profile development (Fig. 8–5). From the McFadden and Tinsley (1985) simulations, it is clear that a "wetter" Pleistocene does not necessarily imply a greater rainfall; a large change in temperature, per se, is sufficient for polygenetic profile development. These Pleistocene simulations (Marion et al., 1985; McFadden & Tinsley, 1985) are good examples of how simulation models may be useful in evaluating paleo-processes in soils.

Spatial and Temporal Variability

Spatial and temporal variability are important factors in simulation models because variability affects both model parameterization and validation. Several other papers in this symposium deal explicitly with variability (Addiscott, 1994; Slater et al., 1994; Wagenet et al., 1994; Wilding et al., 1994). At times it is possible to cope with variability by using averages, at other times it is necessary to explicitly recognize variability and use stochastic models.

Most $CaCO_3$ formation models explicitly allow for profile (spatial) variability in pH and P_{CO2} but not for seasonal (temporal) variability (Table 8–1). Field studies in deserts show clearly that there is significant seasonal variation in P_{CO2} (Parada et al., 1983; Amundson et al., 1988; Quade et al., 1989; Terhune & Harden, 1991). Given the critical role of P_{CO2} in controlling soil pH and Ca concentrations in carbonate systems (Fig. 8–2), both spatial and temporal variability in P_{CO2} should be explicitly incorporated into soil $CaCO_3$ models.

Most calcareous soils are supersaturated with respect to pure macroscopic calcite (Marion & Babcock, 1977; Suarez, 1977; Inskeep & Bloom, 1986; Amrhein & Suarez, 1987; Marion et al., 1990; Suarez et al., 1992). The degree of supersaturation may vary widely between and within soils, with profound effects on calculated pH. For example, there is an approximately 0.3 pH unit difference between the Alaska and Mojave soils (Fig. 8–3) which is caused primarily by an order-of-magnitude difference in $CaCO_3$ solubility (K_{cal}). For a desert soil near Las Cruces, New Mexico, horizontal variability is shown by the error bars in Fig. 8–6, while the bars, per se, reflect vertical variability within the soil profile. The theoretical solubilities of Ca at P_{CO2} = 0.035 atm (0.00354 MPa) and $[SO_4^{2-}]$ = 1 $\times 10^{-5}$ M for K_{cal} = 9.33×10^{-9} (A horizon) and K_{cal} = 5.50×10^{-9} (B_{k2} horizon) (Fig. 8–6) are 7.27×10^{-4} and 6.01×10^{-4}, respectively (Eq. [2]). Calcium should theoretically be 21% more soluble in the A horizon than in the B_{k2} horizon. These calculations demonstrate the importance of selecting appropriate equilibrium constants in simulation models.

Most parameterizations of rainfall in models of $CaCO_3$ formation are deterministic, using the same average rainfall pattern year-after-year, and these

CHIHUAHUAN DESERT SOIL

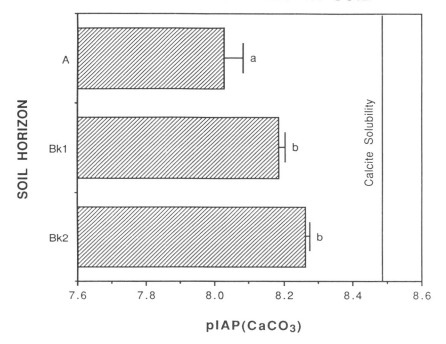

Fig. 8–6.   The mean (±1 SE) pIAP(CaCO$_3$) as a function of soil horizon for a Chihuahuan Desert soil. A different lowercase letter implies a statistically significant difference (adapted with permission from Marion et al., 1990).

models frequently use monthly time steps (Table 8–1). Such models are likely to obscure both the effects of extreme rainfall events within years and extreme rainfall years within centuries.

There were "apparent" anomalies in the predicted depth of the CaCO$_3$ layer in the CALDEP model (Fig. 8–5). For example, the mean depth was similar for Yuma and Albuquerque despite a large difference in mean annual rainfall (8.5 vs. 21.1 cm). In New Mexico, Roswell showed a much deeper mean depth of CaCO$_3$ deposition (54 cm) than Clayton (40 cm) despite a greater mean annual rainfall at Clayton (38 cm) than at Roswell (32 cm). These anomalies were attributed to the greater frequency of extreme rainfall events within years for the Yuma and Roswell sites compared to the Albuquerque and Clayton sites (Marion et al., 1985).

The mean annual rainfall (±1 standard deviation [SD]) for Tucson, Arizona, is 28.4 ± 7.5 cm (Marion et al., 1985). This variation implies that about 2.5% of the time, one could expect annual rainfall ≤13.4 cm, and 2.5% of the time one could expect annual rainfall ≥43.4 cm (mean ±2 SD). Extremely wet years could strongly influence the depth of salt movement in soils.

We compared the stochastic rainfall model used to simulate salt movement in soils (Fig. 8–4) with a deterministic model which used the same rainfall pattern year-after-year. Rain dates and rainfall amounts for the deterministic model were taken from the 1980 weather record for Tucson. During that year, 31.1 cm of rain fell in Tucson. By eliminating 6 of 51 rain dates, we reduced total annual rainfall to 28.4 cm, which is the long-term average annual rainfall for Tucson. This amended rainfall pattern was then used to parameterize the deterministic model.

All of the salt precipitation occurred in the 30- to 50-cm soil layers using the deterministic model (Fig. 8–7). Apparently there was little water movement past the 50-cm depth in these simulations. Gypsum precipitated in the deeper soil layer (40–50 cm) as one might expect given its greater solubility. Compared to the stochastic model (Fig. 8–4B), there was a shallower depth of salt precipitation for the deterministic model (Fig. 8–7). For example, gypsum precipitation occurred primarily in the 90- to 100-cm layer for the stochastic model but at 40 to 50 cm for the deterministic model.

However, comparing a stochastic rainfall model with a deterministic rainfall model is exceedingly tricky. Not only must both models give the same average annual rainfall, but individual events within a year for the deterministic model also must reproduce the long-term average magnitude and frequency of these events at a given site. The maximum rainfall event used in the deterministic model, based on actual weather records for Tucson in 1980, was 2.7 cm. The stochastic model allows 1.7% of winter rainfall amounts to vary between 3.0 and 7.5

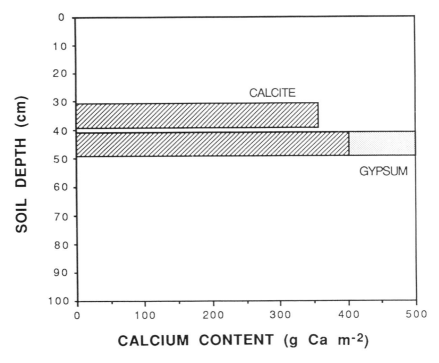

Fig. 8–7. Simulation (1000 yr) of calcite and gypsum formation using a deterministic rainfall model.

cm, thus occasionally producing more intense storms than allowed in the deterministic model.

Despite these limitations, the comparison of the stochastic (Fig. 8–4B) and the deterministic (Fig. 8–7) models demonstrates the relative difference in soil properties produced by these two models. In general, a deterministic model will have a shallower zone of salt deposition than a stochastic model. The reason is quite simple. Two years with 13.4 and 43.4 cm of annual rainfall are not equivalent to 2 yr with 28.4 cm of annual rainfall in terms of water and salt movement in soils. Rainfall patterns are an example of a process that can not be adequately described by averaging. To model salt movement in soils, one should use a stochastic rainfall model which explicitly recognizes extreme rainfall events within years and extreme rainfall years within centuries.

SUMMARY

These four applications were selected in order to demonstrate how simulation models may be useful to studies of soil chemistry and pedogenesis. The pH-sulfate case exemplified integration of empirical and theoretical knowledge that identified a gap in our understanding (how pH and sulfate concentrations are related in calcareous soils). This improved understanding was then incorporated into the CALGYP model. Predicting calcite-gypsum formation in soils is an inherently complex problem that necessitates a multidisciplinary approach integrating climatic, edaphic, biotic, and mathematical factors. Under paleoprocesses, we examined several hypotheses about Pleistocene climates, only a few of which are compatible with polygenetic development of calcic profiles in the desert Southwest. Finally, we examined how variability in soils affects model parameterization and validation. Some parameters and processes can be dealt with by averaging, while others, such as rainfall patterns, require stochastic models as an explicit recognition of variability.

In soil science, a strong synergism exists between empirical studies and theoretical models, both of which contribute to a multidisciplinary approach in solving complex problems and ultimately to a better integrative understanding of soil-forming processes.

REFERENCES

Addiscott, T.M. 1994. Simulation, prediction, foretelling or prophecy? Some thoughts on pedogenetic modeling. p. 1–15. *In* R.B. Bryant and R. Arnold (ed.) Quantitative modeling of soil forming processes. SSSA Spec. Publ. 39. SSSA, Madison, WI.

Amrhein, C., and D.L. Suarez. 1987. Calcite supersaturation in soils as a result of organic matter mineralization. Soil Sci. Soc. Am. J. 51:932–937.

Amundson, R.G., O.A. Chadwick, J.M. Sowers, and H.E. Doner. 1988. Relationship between climate and vegetation and the stable carbon isotope chemistry of soils in the eastern Mojave desert, Nevada. Quat. Res. 29:245–254.

Amundson, R.G., and L.J. Lund. 1987. The stable isotope chemistry of a native and irrigated Typic Natrargid in the San Joaquin Valley of California. Soil Sci. Soc. Am. J. 51:761–767.

Arkley, R.J. 1963. Calculation of carbonate and water movement in soil from climatic data. Soil Sci. 96:239–248.

Birkeland, P.W. 1984. Soils and geomorphology. Oxford Univ. Press, New York.

Davies, C.W. 1962. Ion association. Butterworths, London.

Dudley, L.M., R.J. Wagenet, and J.J. Jurinak. 1981. Description of soil chemistry during transient solute transport. Water Resour. Res. 17:1498–1504.

Dutt, G.R., M.J. Shaffer, and W.J. Moore. 1972. Computer simulation model of dynamic bio-physiochemical processes in soils. Arizona Agric. Exp. Stn. Tech. Bull. 196.

Galloway, R.W. 1983. Full-glacial southwestern United States: Mild and wet or cold and dry. Quat. Res. 19:236–248.

Gile, L.H., J.W. Hawley, and R.B. Grossman. 1981. Soils and geomorphology in the Basin and Range area of southern New Mexico-Guidebook to the desert project. Memoir 39. New Mexico Bur. Mines Miner. Resour. Socorro, NM.

Harden, J.W., E.M. Taylor, M.C. Reheis, and L.D. McFadden. 1991. Calcic, gypsic, and siliceous soil chronosequences in arid and semiarid environments. p. 1–16. In W.D. Nettleton (ed.) Occurrence, characteristics, and genesis of carbonate, gypsum, and silica accumulations in soils. SSSA Spec. Publ. 26. SSSA, Madison, WI.

Hendricks, D.M. 1991. Genesis and classification of arid regions soils. p. 33–79. In J. Skujins (ed.) Semiarid lands and deserts: Soil resource and reclamation. Marcel Dekker, Inc., New York.

Inskeep, W.P., and P.R. Bloom. 1986. Calcium carbonate supersaturation in soil solutions of Calciaquolls. Soil Sci. Soc. Am. J. 50:1431–1437.

Knuteson, J.A., J.L. Richardson, D.D. Patterson, and L. Prunty. 1989. Pedogenic carbonates in a Calciaquoll associated with a recharge wetland. Soil Sci. Soc. Am. J. 53:495–499.

Lindsay, W.L. 1979. Chemical equilibria in soils. John Wiley & Sons, New York.

Machette, M.N. 1985. Calcic soils of the southwestern United States. p. 1–21. In D.G. Weide (ed.) Soils and quaternary geology of the southwestern United States. Geol. Soc. Am. Spec. Pap. 203. Geol. Soc. Am., Boulder, CO.

Marion, G.M. 1989. Correlation between long-term pedogenic $CaCO_3$ formation rate and modern precipitation in deserts of the American Southwest. Quat. Res. 32:291–295.

Marion, G.M., and K.L. Babcock. 1977. The solubilities of carbonates and phosphates in calcareous soil suspensions. Soil Sci. Soc. Am. J. 41:724–728.

Marion, G.M., D.S. Introne, and K. Van Cleve. 1991. The stable isotope geochemistry of $CaCO_3$ on the Tanana River floodplain of interior Alaska: Composition and mechanisms of formation. Chem. Geol. (Isotope Geosci.) 86:97–110.

Marion, G.M., W.H. Schlesinger, and P.J. Fonteyn. 1985. CALDEP: A regional model for soil $CaCO_3$ (caliche) deposition in southwestern deserts. Soil Sci. 139:468–481.

Marion, G.M., W.H. Schlesinger, and P.J. Fonteyn. 1990. Spatial variability of $CaCO_3$ solubility in a Chihuahuan Desert soil. Arid Soil Res. Rehab. 4:181–191.

Marion, G.M., K. Van Cleve, C.T. Dyrness, and C.H. Black. 1993. The soil chemical environment along a forest primary successional sequence on the Tanana River floodplain, interior Alaska. Can. J. For. Res. 23:914–922.

Mayer, L., L.D. McFadden, and J.W. Harden. 1988. The distribution of calcium carbonate: A model. Geology 16:303–306.

McFadden, L.D. 1982. The impacts of temporal and spatial climatic changes on alluvial soil genesis in southern California. Ph.D. diss. Univ. Arizona, Tucson, AZ (Diss. Abstr. 83-04724).

McFadden, L.D., R.G. Amundson, and O.A. Chadwick. 1991. Numerical modeling, chemical, and isotopic studies of carbonate accumulation in soils of arid regions. p. 17–35. In W.D. Nettleton (ed.) Occurrence, characteristics, and genesis of carbonate, gypsum, and silica accumulations in soils. SSSA Spec. Publ. 26. SSSA, Madison, WI.

McFadden, L.D., and J.C. Tinsley. 1985. Rate and depth of pedogenic-carbonate accumulation in soils: Formulation and testing of a compartment model. p. 23–41. In D.G. Weide (ed.) Soils and quaternary geology of the southwestern United States. Geol. Soc. Am. Spec. Pap. 203, Geol. Soc. Am., Boulder, CO.

Parada, C.B., A. Long, and S.N. Davis. 1983. Stable-isotopic composition of soil carbon dioxide in the Tucson Basin, Arizona, U.S.A. Isotope Geosci. 1:219–236.

Quade, J., T.E. Cerling, and J.R. Bowman. 1989. Systematic variations in the carbon and oxygen isotopic composition of pedogenic carbonate along elevation transects in the southern Great Basin, United States. Geol. Soc. Am. Bull. 101:464–475.

Reheis, M.C. 1987. Gypsic soils on the Kane alluvial fans, Big Horn County, Wyoming. U.S. Geol. Surv. Bull. 1590-C. U.S. Geol. Serv., Denver, CO.

Robbins, C.W., R.J. Wagenet, and J.J. Jurinak. 1980. A combined salt transport chemical equilibrium model for calcareous and gypsiferous soils. Soil Sci. Soc. Am. J. 44:1191–1194.

Rogers, R.J. 1980. A numerical model for simulating pedogenesis in semiarid regions. Ph.D. diss. Univ. Utah, Salt Lake City (Diss. Abstr. 80-22642).

Schlesinger, W.H. 1985. The formation of caliche in soils of the Mojave Desert, California. Geochim. Cosmochim. Acta 49:57–66.

Simunek, J., and D.L. Suarez. 1992. Two-dimensional model of unsaturated water flow and equilibrium carbonate chemistry. p. 228. *In* Agronomy abstracts. ASA, Madison, WI.

Slater, B.J., K. McSweeney, A.B. McBratney, S.J. Ventura, and B.J. Irvin. 1994. A spatial framework for integrating soil-landscape and pedogenic models. p. 169–185. *In* R.B. Bryant and R. Arnold. (ed.) Quantitative modeling of soil forming processes. SSSA Spec. Publ. 39. SSSA, Madison, WI.

Suarez, D.L. 1977. Ion activity products of calcium carbonate in waters below the root zone. Soil Sci. Soc. Am. J. 41:310–315.

Suarez, D.L., J.D. Wood, and I. Ibrahim. 1992. Reevaluation of calcite supersaturation in soils. Soil Sci. Soc. Am. J. 56:1776–1784.

Terhune, C.L., and J.W. Harden. 1991. Seasonal variations of carbon dioxide concentrations in stony, coarse-textured desert soils of southern Nevada, USA. Soil Sci. 151:417–429.

Van Devender, T.R., and W.G. Spaulding. 1979. Development of vegetation and climate in the southwestern United States. Science (Washington, DC) 204:701–710.

Wagenet, R.J., J. Bouma, and J.L. Hutson. 1994. Modeling water and chemical fluxes as driving forces of pedogenesis. p. 17–35. *In* R.B. Bryant and R. Arnold (ed.) Quantitative modeling of soil forming processes. SSSA Spec. Publ. 39. SSSA, Madison, WI. (This volume.)

Wells, P.V. 1966. Late Pleistocene degree of pluvial climatic change in the Chihuahuan Desert. Science (Washington, DC) 153:970–975.

Wells, P.V. 1979. An equable glacipluvial in the West: Pleniglacial evidence of increased precipitation on a gradient from the Great Basin to the Sonoran and Chihuahuan Deserts. Quat. Res. 12:311–325.

Wilding, L.P., J. Bouma, and D. Goss. 1993. Impact of spatial variability on interpretive modeling. p. 61–75. *In* R.B. Bryant and R. Arnold (ed.) Quantitative modeling of soil forming processes. SSSA Spec. Publ. 39. SSSA, Madison, WI. (This volume.)

Zelichenko, Y.N., and S.A. Sokolenko. 1982. Calculation of the rate of development of new carbonate formations in soils. Sov. Soil Sci. 14 (1):111–117.

9

A General Model for Soil Organic Matter Dynamics: Sensitivity to Litter Chemistry, Texture and Management

William J. Parton and Dennis S. Ojima

Colorado State University
Fort Collins, Colorado

C. Vernon Cole

USDA-ARS
Colorado State University
Fort Collins, Colorado

David S. Schimel

National Center for Atmospheric Research
Boulder, Colorado

Different conceptual and simulation models of soil organic matter (SOM) have been developed during the last 20 yr. Jenkinson and Rayners (1977) developed the first widely used SOM model. This model divided soil C into active, slow and passive pools that have different turnover times (1 yr, 30 yr, and 1500 yr, respectively). This conceptual framework has been used in many of the recently developed SOM models (van Veen & Paul, 1981; Parton et al., 1987; Jenkinson, 1990). van Veen and Paul (1981) improved on Jenkinson and Rayners (1977) model by including concepts of physical and chemical protection, and factors such as soil erosion and soil cultivation. Parton et al. (1987) added the impact of soil texture on SOM dynamics and developed generalized nutrient cycling submodels which simultaneously simulate soil C and inorganic and organic N, P and S dynamics (Parton et al., 1988). Jenkinson's most recent model (Jenkinson, 1990) also includes soil physical protection by clay. Many of the concepts used in the different SOM models have been incorporated into the nutrient cycling models that are linked to crop plant production models [Erosion-Productivity Impact Calculator (EPIC), Williams et al., 1985; Nitrogen-Tillage-Residue Management (NTRM), Shaffer, 1985; DeNitrification and DeComposition Model (DNDC), Li et al., 1992]. Some of the more recent models (e.g., CENTURY, Parton et al., 1987; DNDC, Li et al., 1992) include the impact of soil texture on SOM dynamics and are used to simulate detailed nutrient cycling dynamics and trace gas fluxes.

Martin and Haider (1986) reviewed the literature on the impact of soil clay content, plant lignin and N content, and soil mineralogy on soil C stabilization. They suggest that plant derived lignin and microbial melanins play an important role in soil C stabilization, that soil C stabilization increases with the amount of clay with soil, that short-term (1-yr) decomposition of plant material are primarily controlled by climatic factors, nutrient availability, and soil pH but have little impact on long-term C stabilization, and that most (≤80%) of the humus fraction of the soil consists of humic acids and polysaccharides. The literature reviewed in this article was used to conceptualize the CENTURY model and derive parameters for the model.

Recent papers have described the global application of the grass production model (Parton et al., 1993). This paper will focus on the conceptual basis for recent changes to CENTURY SOM model and on the impact of soil texture on SOM dynamics. We compare model predictions on the effect of soil texture on total soil C levels and turnover rates of C in different pools, short-term (1–2 yr) dynamics of added plant residue, and loss of soil C due to cultivation with observation. We also summarize the modeling studies of the long-term impact of different crop management practices on soil C and N stabilization at Pendleton, Oregon, site (Parton & Rasmussen, 1994), Sidney, Nebraska (Metherell, 1992; Metherell et al., 1994), and in Sweden (Paustian et al., 1992). We will show how the model was used to interpret observed long-term soil C data sets, to highlight uncertainties in our understanding about SOM dynamics and to suggest research topics that are critical for advancing our knowledge about SOM.

CENTURY MODEL DESCRIPTION

A detailed description of the structure of the CENTURY model and the concepts embodied in the model are presented in Parton et al. (1987) and recent modifications are described by Parton et al. (1993). The compartment diagram (Fig. 9–1) shows that surface and root litter material are represented separately, incoming plant material is divided into structural and metabolic material, surface microbes are separately modeled and that soil SOM pools include the active (microbe), slow and passive pools. The active pool consists of live soil microbes and microbial products with a short turnover time (residence time of -1 yr), the slow pool includes plant and microbial products that are physically protected or biologically resistant to decomposition (residence time of 10's of yr), while the passive pool can be chemically recalcitrant or physically protected (residence times of 100's–1000's of yr). The actual turnover time of these pools is a function of the abiotic decomposition factor which is controlled by the monthly soil temperature and ratio of rainfall plus available water to potential evapotranspiration rate (Fig. 9–2). In the SOM submodel, decomposing plant residue is split between structural and metabolic plant material as a function of the initial lignin-to-N ratio. The structural material is assumed to contain cellulose and all of the

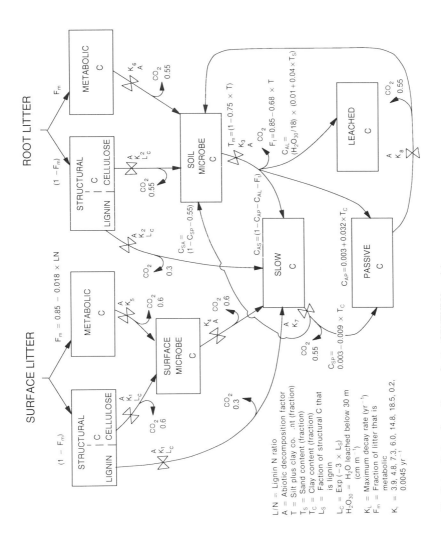

Fig. 9–1. Carbon flow diagram for the CENTURY model.

lignin (Fig. 9–1). The decomposition rate of structural material is a function of the lignin content in the structural material. The lignin fraction is transferred directly into slow SOM as structural material decomposes. Active SOM turnover rate is a function of soil texture, and formation of passive SOM is a function of the clay content. Recent major revisions to the original version of the CENTURY model (Parton et al., 1987) include (i) adding the impact of clay on formation of passive SOM, (ii) improving the surface litter decomposition model, (iii) revising the equations for the impact of moisture on decomposition (including the effect of anaerobic conditions and stored soil water, Fig. 9–2), and (iv) representing leaching of soluble SOM.

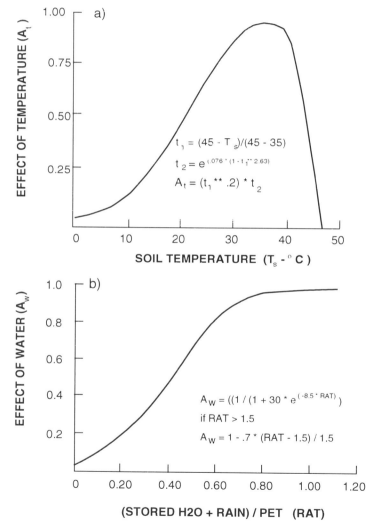

Fig. 9–2.   The effect of soil temperature (*a*) and soil water (*b*) on decomposition. The equations for the curves are presented on the figure.

Decomposition of each state variable is calculated using the following equations

$$\frac{dC_1}{dt} = K_1 L_C A C_I \qquad I = 1, 2 \qquad [1]$$

$$\frac{dC_1}{dt} = K_I A T_m C_I \qquad I = 3 \qquad [2]$$

$$\frac{dC_1}{dt} = K_1 A C_I \qquad I = 4, 5, 6, 7, 8 \qquad [3]$$

$$T_m = (1 - 0.75T) \qquad [4]$$

$$L_C = e^{(-3L_s)} \qquad [5]$$

where C_I = the C in state variable; I = 1,2,3,4,5,6,7,8, for surface and soil plant residue structural material, active SOM, surface microbes, surface and soil plant residue metabolic material, and slow and passive SOM fractions; K_I = maximum decomposition rate (yr⁻¹) parameters for the Ith state variables (K_I = 3.9, 4.9, 7.3, 6.0, 14.8, 18.5, 0.20, 0.0045 yr⁻¹); A is the combined abiotic impact of soil moisture and soil temperature on decomposition (product of the soil moisture and temperature terms, see Fig. 9–2); T_m is the effect of soil texture (T) on active SOM turnover [i.e., $1 - 0.75 \times (T)$, where T is equal to silt + clay fraction]; and L_C is the impact of lignin content of structural material (L_s) on structural decomposition. The multiplication approach was used for the combined effect of temperature and moisture on decomposition. Many models use the multiplication approach (Hunt, 1977; McGill et al., 1981), however, it is unclear which is the best approach from a theoretical point of view.

The model assumes that C decomposition flows are associated with microbial activity and that microbial respiration occurs for each of these flows. The fraction of C lost due to microbial respiration with each C flow is shown in Fig. 9–1 next to the CO_2 arrows. Carbon leaving the active SOM box (surface microbes are separate) is divided into four different flows which include microbial respiration, leaching of soluble organic C, and stabilization of C in the slow and passive pools. Equations [6–9] are used to calculate the fraction of the total C flow out of active SOM that is allocated to those four flows.

$$F_T = 0.85 - 0.68T \qquad [6]$$

$$C_{AL} = (H_2O_{30}/18)(0.01 + 0.04T_s) \qquad [7]$$

$$C_{AP} = 0.003 + 0.032T_c \qquad [8]$$

$$C_{AS} = (1 - C_{AP} - C_{AL} - F_T) \qquad [9]$$

where, F_T is the fraction of C lost due to microbial respiration, C_{AL} is the fraction lost due to organic leaching, C_{AP} is the fraction allocated to passive SOM, C_{AS} is the fraction sent to the slow pool, T is equal to the silt plus clay content (fraction),

H_2O_{30} is the monthly water leached below the 30-cm soil depth (cm mo^{-1}), T_s is the sand content (fraction), and T_c is the soil clay content (fraction). Carbon that flows out of the slow SOM box is allocated to passive SOM and active SOM using Eq. [10] and [11]

$$C_{SP} = 0.003 - 0.009 \times T_c \qquad [10]$$

$$C_{SA} = \left(1 - C_{SP} - 0.55\right) \qquad [11]$$

where C_{SP} is the fraction of C allocated to passive pool, and C_{SA} is the fraction allocated to the active pool (55% of the C is lost due to microbial respiration).

LITTER DECOMPOSITION

The CENTURY model simulates the decomposition of both surface and soil litter material. The surface litter decomposition model was recently revised (Vitousek et al., 1994; Parton et al., 1994) to include a surface litter microbe pool (Fig. 9–1) and tested extensively using litter decomposition data from Hawaii. The data included information from a common site where the litter quality was varied experimentally (different N and lignin content) and data from many sites where a common litter type was placed in different environmental conditions (i.e., varying levels of temperature, water, and soil nutrients). A comparison of observed and simulated results of C remaining in litter placed in the field for up to 2 yr show that the model agreed with litter decomposition data across environmental gradients (mean annual air temperature ranged from 10–24°C, the annual precipitation ranged from 500–3800 mm yr^{-1} and the annual actual evapotranspiration rates ranged from 500–1640 mm yr^{-1}; $r^2 = 0.83$, Fig. 9–3a). Model results from a study using litter with fixed litter quality and across a litter quality gradient also compared well to observed data (leaf litter lignin content varied from 13–32%, N contents varied from 0.3–0.5% and the L/N ratios varied from 25–90, $r^2 = 0.85$, Fig. 9–3b) under the same environmental conditions as the previous analysis. A detailed comparison of the simulated N and P release from decomposing litter (data not presented here) shows that N and P parameterization presented in the Vitousek et al. (1994) paper tends to overestimate N and P release from litter. In general, there are larger differences in N and P release (as compared to C release) between litter types. Recent results suggest that you need to include leaching of organic material out of the litter in order to adequately simulate N and P dynamics.

The litter decomposition model assumes that plant material is composed of readily decomposable material (metabolic) and material that is resistant to decomposition (structural). The model calculates the split between structural and metabolic material as a function of the initial lignin to N ratio of the plant material. Chemical analysis of 30 different types of plant material (root and leaf litter) were used to test this assumption. This initial litter chemistry data represents plant material from diverse temperate and tropical grasslands, forests and agroecosystems which were collected as part of a long-term intersite decomposition

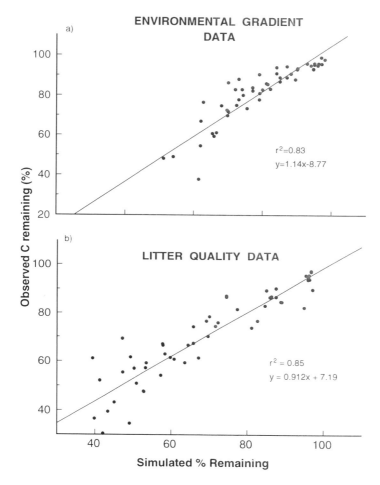

Fig. 9–3. Comparison of the observed vs. simulated percentage C remaining for (*a*) litter decomposition along environmental gradients in Hawaii (*b*) and for a common site in Hawaii with different litter types (initial N and lignin contents). This figure is redrawn from data presented by Vitousek et al. (1994, p. 21).

study (LIDET, 1993, unpublished data). Lignin contents of the different plant material ranged from 5 to 35%, while the N content ranged from 0.3 to 2.5%. The chemical extraction techniques of Ryan et al. (1990) were used to separate cellulose (acid soluble carbohydrates), lignin and water soluble extractable material (e.g., alcohols, sugars, soluble phenols, amino acids). Cellulose and lignin fractions are structural components, while the water soluble extractive fraction is equivalent to CENTURY's metabolic fraction. Figure 9–4 shows that the cellulose plus lignin fraction increases linearly with increasing L/N ratio and the water soluble fraction decreases with increasing L/N ratio, as predicted by Eq. [6] used in the CENTURY model (Fig. 9–1). Despite the general pattern shown in Fig.

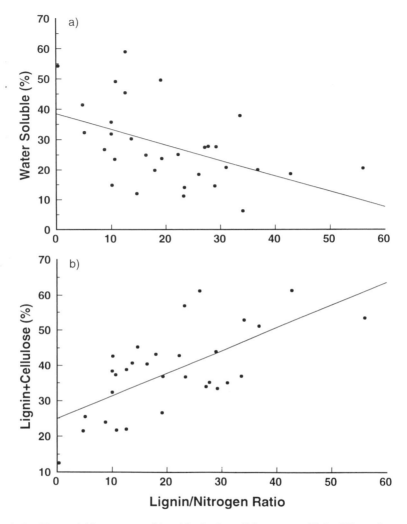

Fig. 9–4. Water soluble component (*a*) and lignin plus cellulose content (*b*) for different litter types as a function of the initial litter lignin to N ratio. Best fit linear equation is shown in the figure.

9–4(a,b), the scatter of points indicate that individual species have quite different partitioning of organic matter between metabolic and structural material than the general equation would predict.

Litter decomposition data set from Sweden (Berg & Ekbohm, 1991; Berg & McClaugherty, 1989) also was used to test the surface litter decomposition model. Berg and Ekbohm's (1990) data (Fig. 9–5b) for three different litter types [Alder (genus *Alnus*), Scots pine (*Pinus sylvestris* L.) and lodgepole pine (Pinus contorta Douglas ex London)] which have similar lignin contents (26–34%) and different initial N contents (3% for Alder, 1.5% for Scots pine and 0.7% for

Lodgepole pine) show that initial C loss was fastest for high N content material (i.e., Alder material). This pattern was reversed after 2 yr of decomposition with faster C loss in the low N litter materials (Scots and Lodgepole pine) during the latter phases of decomposition. The net result after 4 yr is for more C loss from low N litter than high N litter. The model CENTURY simulation of these three litter types (Fig. 9–5a) shows that C losses are highest initially for the high N litter (Fig. 9–5a), however, the model underestimates the initial C loss from the low N content litter and does not simulate the reduction in decomposition rate of the high N content litter accounting for the overall lower total C loss of the high N content Alder litter after 4 yr. The most likely explanation for underestimate of C loss for low N litter is that the model underestimated the fraction of C in metabolic C pool.

Berg and Ekbohm (1990) suggested that the reason for the higher long-term C stabilization (lower C loss) for initial high N litter was that the stabilized

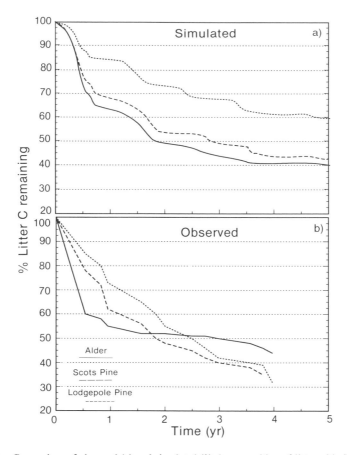

Fig. 9–5. Comparison of observed (*a*) and simulated (*b*) decomposition of litter with different initial N contents. The litter was placed in a Scots pine forest from 1983 to 1986 at a site in Sweden (Berg & Ekbohm, 1991).

decomposition products (slow SOM) have lower decomposition rates. We tested these results using the CENTURY model, and showed that much lower decomposition rates are required for stabilized microbial products (slow SOM) created from initial high N compounds in order to duplicate the observed litter decomposition patterns. The formation of resistant slow SOM from high initial N content material was noted previously by Berg and Wessen (1984), however, the degree of the impact is much larger in the Berg and Ekbohm (1990) study. The implications of these studies to further revisions to CENTURY are not well defined, however, these studies suggest that decomposition rates of slow SOM may need to be a function of either the initial N content of decomposing litter or the fraction of the structural plant material that is lignin (lower decomposition rates for higher ratios).

SOIL TEXTURAL EFFECT ON SOIL ORGANIC MATTER DYNAMICS

In CENTURY, soil texture influences the turnover rate of active SOM (microbial biomass and products) and the stabilization efficiency of slow SOM. The turnover rate of active SOM decreases linearly with increasing silt plus clay soil content (Eq. [4]), and the stabilization efficiency of slow SOM increases (Eq. [6]) with increasing silt plus clay content (Parton et al., 1987). These assumptions are based on long-term (1–5 yr) soil incubation data where different types of plant material were added to soils with different soil textures (Sorenson, 1981) and other observations by Van Veen et al. (1984, 1985), Gregorich et al. (1991), and Ladd et al. (1981). These assumptions were recently tested using soil incubation data (Amato & Ladd, 1992) where ^{14}C labeled plant material (alfalfa, *Medicago sativa* L.) was added to Australian soils. We simulated Ladd et al.'s laboratory experiments using a microcosm version (laboratory incubation version) of the CENTURY model by adding alfalfa plant material to soils that had different clay and sand contents (with silt content constant at 15%). We compared the observed and simulated total soil C and microbial biomass (active SOM in the model) after a 16-mo incubation at 25°C. Both the model results and data for total C remaining (Fig. 9–6) show that soil C stabilization increases linearly with increasing clay content. The comparison of simulated (active SOM) and observed microbial biomass show that microbial biomass doubles as clay content increases from 10 to 50%. The model underestimated active (microbial biomass) SOM levels (values are 30–40% of the observed). The reason for this discrepancy is unclear, however, it suggests that active SOM turnover rates may be too high. More data is needed to test whether the maximum active decomposition rate needs to be changed.

In CENTURY, the formation of passive SOM is controlled by clay content. The flow of C from active SOM and slow SOM to passive SOM are a linear function of the soil clay content (see Eq. [10] and [11]). This formulation is based on the assumption that clay materials are the primary particulates that stabilize passive SOM. Greenland (1965) and Giovannini and Sequi (1976) suggest that clay and metal ions on clay surfaces are involved in the stabilization of humic colloids. Turchenek and Oades (1979) showed that 37 to 40% of organic C was recovered in the clay fraction. The interactions of humic substances and clays are influenced

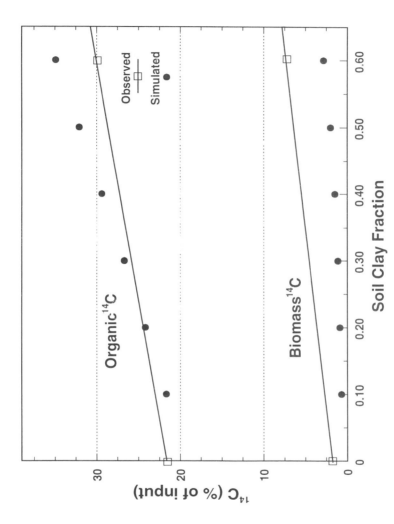

Fig. 9–6.   Comparison of observed and simulated total ¹⁴C and microbial biomass ¹⁴C remaining after a 16-mo incubation of alfalfa litter at 25°C. The data were presented by Amato and Ladd (1992).

by pH, the type of exchangeable cations, specific area of the clays, charge density, and surface electrical charge of the clay, however, these specific mechanisms are not directly simulated (Theng, 1980; Tate & Theng, 1980; Oades et al., 1989). There is an extensive body of literature that support the effect of clay content on SOM stabilization, however, it is difficult to use available data to quantify the flows of C to passive SOM. The parameters used in Eq. [10] and [11] and the maximum decay rate of passive SOM were determined by adjusting the values to match observed soil C levels for different soil textures at two sites in the Great Plains (northeastern Colorado and eastern Kansas). Note that the maximum decay rate of the passive SOM was reduced (as compared to Parton et al., 1987) as part of the process of including the effect of soil clay content on the formation of passive SOM.

Most work on the relationships between clay content and SOM has been conducted in soils with permanent charge mineralogies (Huang & Schnitzer, 1986). A variety of authors (Tate & Theng, 1980; Oades et al., 1989; Goh, 1980) suggest that the impact of clay on C stabilization will change as a function of clay mineralogy. We are presently attempting to quantify the effects of mineralogy using long-term soil incubation data for temperate and tropical soils with different clay mineralogy (smectitic, kaolinitic, oxidic and alleophanic) and across a range of soil textures. Preliminary results suggest that the size of the passive pool is larger for oxidic and alleophanic mineralogy in comparison to kaolinitic and smectitic ones.

We ran the model to steady-state conditions for a tallgrass prairie site in Russia for eight different soil textures (Fig. 9–7). Total SOM C levels (Fig. 9–7a) increased with increasing clay content, and the fraction of total SOM C in the passive fraction (Fig. 9–7b) and the slow + passive SOM turnover time (Fig. 9–7c) increased linearly with increasing clay content. Increasing the sand content for a given clay content (decreasing silt content) caused total soil C to decrease rapidly and soil C turnover time to increase. Model results showed that most of the increase in total soil C with increasing clay content resulted from increases in passive soil C. The fraction of total C in passive SOM increased from 30 to over 50% as clay content increased from 10 to 50%.

Data useful in testing model calculations of texture effects are presented by Burke et al. (1989) and Becker-Heidman (1989). Becker-Heidman measured the ^{14}C mean residence time for soils in Germany with different clay contents. The simulated residence times of the slow plus passive SOM (0–20-cm depth) are similar to Becker-Heidman's observed ^{14}C mean residence times (MRT) at the 10-cm depth (Fig. 9–8a,b). The simulated soil turnover time (0–20 cm) and the observed ^{14}C MRT at 20- and 40-cm depths increased with increasing clay content (data for the 10-cm depth were highly variable and did not show any pattern with texture). This is an approximate comparison, since our simulations were for a site with similar but not identical climate to the German study area. The ^{14}C MRT's of the German soils increase rapidly with depth from less than 144 yr at 10 cm to over 1000 yr at the 40-cm depth (Becker-Heidman, 1989). The biological mechanisms to explain the increase in ^{14}C MRT with depth are not well known, however, model results and data suggest an increase in passive SOM with increasing clay content.

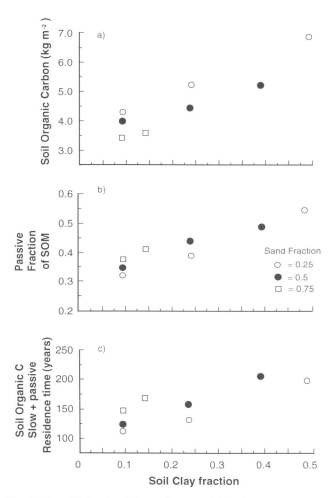

Fig. 9–7. Simulated equilibrium levels for a tallgrass prairie in Eastern Russia (Khomtov) for: (*a*) total SOC, (*b*) fraction of total C in the passive pool, and (*c*) slow and passive residence time.

Cultivation Effects

Model results (Fig. 9–7) show that the amount of passive SOM and the fraction of total SOM in passive SOM increase with soil clay content. This suggests that the fraction of total soil C losses due to cultivation should decrease with increasing soil clay content, since more of the total soil C is in a very resistant fraction in high clay soils. Comparison of results from Burke et al.'s (1989) empirical regression model and simulated model predictions for a site in Kansas (Manhattan, KS) (Fig. 9–9) shows that the percentage of C lost due to cultivation decreased linearly with increasing soil clay content in both the empirical and CENTURY models. The model CENTURY tended to overestimate the total C loss relative to Burke's data, however, the significance of

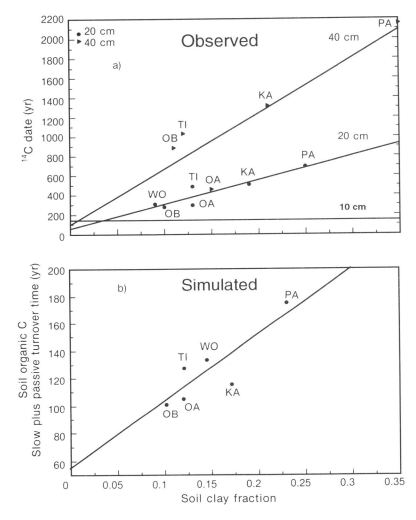

Fig. 9–8.   Observed soil ¹⁴C mean residence time (MRT) for different soils in Germany at the 10-, 20- and 40-cm depth (Becker-Heidman, 1985) (*a*) and simulated MRT (slow plus passive fraction) for the 0- to 20-cm depth at sites with different textures (*b*). Climates were not matched as we had only approximate locations and climate information for ¹⁴C sites. Labels on each point refer to the soils used by Becker-Heidman (1985).

the model overestimate of C loss is unclear since the exact cultivation history for the site is not well known. The diminished losses of SOM in clay soils may be of minimal agronomic benefit, if the remaining SOC is largely in the passive fraction and hence nearly inert with respect to nutrient supply, although the passive SOM does presumably contribute to tilth and erosion resistance.

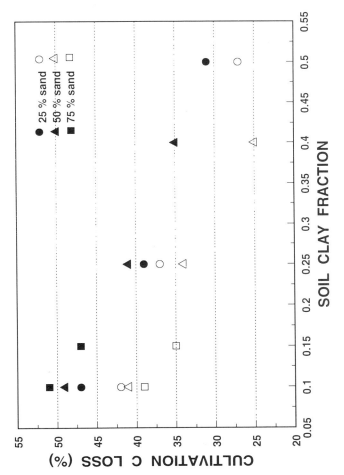

Fig. 9–9. Comparison of observed (Burke et al., 1990) and simulated soil C loss from different soil textures at the Manhattan, Kansas, site. Open symbols are CENTURY results and solid symbols are Burke et al. (1990) results.

Soil Organic Matter Model Testing

The effect of different cultivation practices (no-till, stubble mulch, and conventional plowing) on soil C levels for a wheat-fallow system was analyzed using soil C data from a site in Sidney, Nebraska (Metherell, 1992; Metherell et al., 1994). Both the data and simulation results (Fig. 9–10c) show that soil C losses were least for the no-till plots and highest for the plow treatment after 20 and 30 yr of cultivation. The model did an adequate job of simulating the different cultivation practices, however, the model underestimated plant production for the unfertilized treatments (data not shown). We are uncertain about the cause for this discrepancy, however, N mineralization from the 20- to 50-cm soil layer may be substantial (Parton & Rasmussen, 1994) and is not considered in the CENTURY model. This suggests that the model should represent the dynamics of soil C for deeper soil layers (20–50 cm) in cultivated systems, however, we know little about the dynamics of soil C and nutrients in the 20- to 50-cm layer.

Simulations of soil organic C have been compared to a variety of observed soil C data from grassland systems (Parton et al., 1987; Burke et al., 1989; and Parton et al., 1994, p. 22) and long-term agricultural sites where different types of plant material were added to the soil (Paustian et al., 1992, Parton et al., 1994, p. 22) and different cultivation practices were used (Metherell, 1992; Metherell et al., 1994, p. 21). The comparison of observed and simulated soil C (0–20 cm) data (Fig. 9–11a) for grasslands in the Great Plains (Burke et al., 1989) and soil C data from grasslands around the world (Fig. 9–11b, Parton et al., 1994) show that the model can simulate soil C data for a diverse set of soil textures and climates with r^2's of 0.75 for the Great Plains data and 0.93 for global grassland data. The original version of the CENTURY model (Parton et al., 1987) was used for the comparison with the Great Plains soils data, while the most recent version of the model was used for global grassland data, so the performance difference could be either due to the greater range in the global data or to improvements in the model.

The effect of adding different types and amounts of plant material to the soil (Fig. 9–10a,b) were studied using observed soil C data for 60-yr wheat-fallow plots in Pendleton, Oregon (Parton & Rasmussen, 1994) and 30-yr small grain plots in Sweden (Paustian et al., 1992). Results (Fig. 9–10a,b) show that the model adequately simulated the impact of adding different amounts and types [different lignin contents (8–30%) and N contents (0.4–2%)] of plant material to the soil ($r^2 = 0.8$ and 0.85, respectively for Sweden and Oregon). Inconsistencies in these comparisons between the model results and field data (data not shown) include, underestimated soil C storage for the high N treatments at the Oregon site, underestimated soil N stabilization for the high lignin addition treatments in Sweden, and overestimated soil C loss from the spring burn treatment in Oregon.

CONCLUSIONS

We describe the equations and rationale for the simulation of decomposition and soil organic matter stabilization in the CENTURY model. The CENTURY model has recently been extended and extensively modified to better simulate

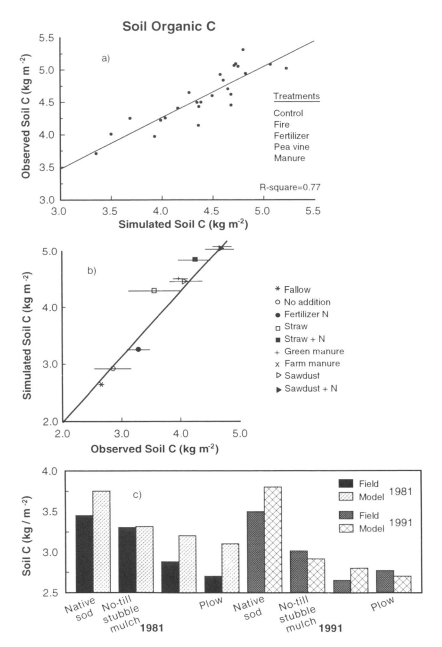

Fig. 9–10. Comparison of observed and simulated soil C for long-term (60-yr) SOM experiments at Pendleton, Oregon (*a*) (redrawn from Parton & Rasmussen, 1994), for long-term (30-yr) SOM experiments in Sweden (*b*) (redrawn from Paustian et al., 1992), and for long-term cultivation experiments at Sidney, Nebraska, (*c*) (redrawn from Metherell, 1992).

Soil C (0-20 cm)

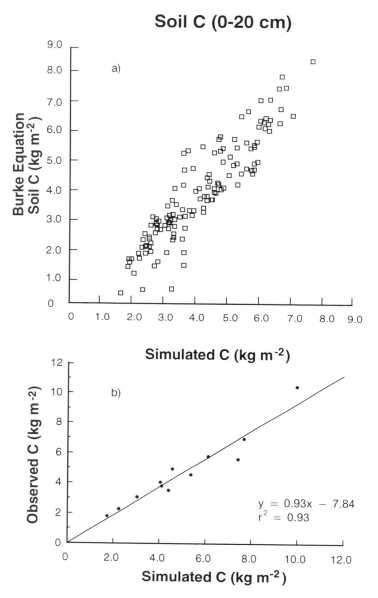

Fig. 9–11.   Comparison of observed vs. simulated soil C (0–20 cm) for sites in the Great Plains (*a*) [redrawn from Burke et al. (1989)] and a global grassland network (*b*) (redrawn from Parton et al., 1993).

litter chemistry and soil texture effects, these changes and model testing are doc-umented. Testing against observations of litter decomposition rates and soil ^{14}C residence times has provided confidence in key, previously untested, aspects of the model. Several important inferences from recent studies include:

1. The widely observed correlation between soil C content and clay content appears to occur primarily as a result of an increase in the fraction of "passive" (residence time >500 yr) organic matter as clay fraction increases. This conclusion is supported by CENTURY model results, ^{14}C residence times observed along texture gradients and effects of texture on C losses in agricultural systems. High clay soils lose approximately the same amount of C under cultivation, however, the fraction of the total C lost is less.

2. Comparison of observed plant chemical fractions to estimates of structural and metabolic fractions from lignin/N ratios shows a generally good cor-relation across wide gradients of litter types, however, the relationship is weakest at high L/N, and considerable deviations from the regression occur on an individual species level.

3. Studies of cultivated systems show that SOC levels and loss rates are influenced by soil texture, tillage management and residue chemistry and addition rate. All of these effects can be simulated with CENTURY, however, problems exist with simulating the effects of added N on C sta-bilization, and with N stabilization when high lignin residues are added to soils.

4. In global (Parton et al., 1993) and regional (Parton et al., 1987) valida-tions of CENTURY, overall agreement between simulated and observed SOC levels is good, with r^2 values of 0.75 to 0.93. The model's ability to simulate C stabilization in natural ecosystems across broad gradients of environmental conditions is excellent. Detailed studies at specific sites, studies of manipulative experiments (litter additions, N addition), and comparisons of observations to specific pools and fluxes in the model has revealed problems in simulating the effects of extreme values of lignin or N in litter, the effects of exogenous N on decomposition and in simulating the release of N and P from litter during decomposition (even though C release is simulated well).

ACKNOWLEDGMENTS

We are grateful to Dr. E. A. Paul for enlightening discussions of earlier drafts of this manuscript. The work was primarily funded by a NASA EOS project no. NAGW-2662 and partially funded by NSF projects (BSR-9108329, BSR-9011659, BSR-9013888, and BSR-9007881). Model development and evaluation was assisted by Alister Metherell, Laura Harding and Becky McKeown. Preparation of the manuscript and graphical material was conducted by Kay McElwain, Becky Techau, Tom Painter, Brian Newkirk, and Michele Nelson. Statistical analysis was conducted by Brian Newkirk. The National Center for Atmospheric Research is sponsored by the National Science Foundation.

REFERENCES

Amato, M., and J.N. Ladd. 1992. Decomposition of ^{14}C-labelled glucose and legume material in soils: Properties influencing the accumulation of organic residue C and microbial biomass C. Soil Biol. Biochem. 24:455–464.

Becker-Heidman, P. 1985. Die Tiefenfunktionen der natürlichen Kohlenstoff-Isotopengehalte von vollständig dünnschichtweise beprobten Parabraunerden und ihre Relation zur Dynamik der organischen Substanz in diesen Böden. Herausgeber: Ver. Förder. Bod., Hamburg, Germany.

Berg, B., and B. Wessen. 1984. Changes in organic-chemical components and ingrowth of fungal mycelium in decomposing birch leaf litter as compared to pine needles. Pedobiologia 26:285–298.

Berg, B., and C. McClaugherty. 1989. Nitrogen and phosphorus release from decomposing litter in relation to the disappearance of lignin. Can. J. Bot. 67:1148–1156.

Berg, B., and G. Ekbohm. 1991. Litter mass-los rates and decomposition patterns in some needle and leaf litter types. Long-term decomposition in a Scots pine forest. VII. Can. J. Bot. 69:1449–1456.

Burke, I.C., C.M. Yonker, W.J. Parton, C.V. Cole, K. Flach, and D.S. Schimel. 1989. Texture, climate, and cultivation effects on soil organic matter context in U.S. grassland soils. Soil Sci. Soc. Am. J. 53:800–805.

Giovannini, G., and P. Sequi. 1976. Iron and aluminum as cementing substances of soil aggregates. II. Changes in the stability of soil aggregates following extractions of iron and aluminum by acetylacetone in a nonpolar solvent. J. Soil Sci. 27:148–153.

Goh, K.M. 1980. Dynamics and stability of organic matter p. 373–393. In B.K.B. Theng (ed.) Soils with variable charge. N.Z. Soc. Soil Sci., Lower Hutt, NZ.

Greenland, D.J. 1965. Interactions between clays and organic compounds in soils. Part 2. Adsorption of soil organic compounds and its effect on soil properties. Soil Fert. 28:415–425.

Gregorich, E.G., R.P. Voroney, and R.G. Kachanoski. 1991. Turnover of carbon through the microbial biomass in soils with different textures. Soil Biol. Biochem. 23:799–805.

Huang, P.M., and M. Schnitzer (ed.). 1986. Interactions of soil minerals with natural organics and microbes. SSSA Spec. Publ. 17. SSSA, Madison, WI.

Hunt, H.W. 1977. A simulation model for decomposition in grasslands. Ecology 58:469–484.

Jenkinson, D.S. 1990. The turnover of organic carbon and nitrogen. Phil. Trans. R. Soc. Lond. B. 329:361–368.

Jenkinson, D.S., and J.H. Rayners. 1977. The turnover of soil organic matter in some of the Rothamsted classical experiments. Soil Sci. 123:298–305.

Ladd, J.N., J.M. Oades and M. Amato. 1981. Microbial biomass formed from ^{14}C ^{15}N-labelled plant material decomposing in soils in the field. Soil Biol. Biochem. 13:119–126.

Li, C., S. Frolking, and T.A. Frolking. 1992. A model of nitrous oxide evolution from soil driven by rainfall events: Model structure and sensitivity. J. Geophys. Res. (Atmospheres) 97:9759–9776.

Martin, J.P., and K. Haider. 1986. Influence of mineral colloids on turnover rates of soil organic carbon. p. 283–304. In P.M. Huang and M. Schnitzer (ed.) Interactions of soil minerals with natural organics and microbes. SSSA Spec. Publ. 17. SSSA, Madison, WI.

McGill, W.B., H.W. Hunt, R.G. Woodmansee, J.O. Reuss, and K.H. Paustian. 1981. Formulation, process controls, parameters and performance of PHOENIX: A model of carbon and nitrogen dynamics in grassland soils. p. 171–191. In M.J. Frissel and J.A. van Veen (ed.) Simulation of nitrogen behaviour of soil-plant systems. Pudoc, Centre for Agric. Publ. Document., Wageningen, the Netherlands.

Metherell, A.K. 1992. Simulation of soil organic matter dynamics and nutrient cycling in agro-ecosystems. Ph.D. diss., Colorado State Univ., Fort Collins (Diss. Abstr. 93-11391).

Metherell, A.K., W.J. Parton, C.A. Cambardella, G.A. Peterson, L.A. Harding, and C.V. Cole. 1994. Simulation of soil organic matter dynamics in dryland wheat-fallow cropping systems. Adv. Soil Sci. (In press.)

Oades, J.M., G.P. Gillman and G. Uehara. 1989. Interactons of soil organic matter and variable-charge clays. p. 69–96. In D.C. Coleman et al. (ed.) Dynamics of soil organic matter in tropical ecosystems. Univ. Hawaii Press, Honolulu.

Parton, W.J., D.S. Schimel, C.V. Cole, and D.S. Ojima. 1987. Analysis of factors controlling soil organic matter levels in Great Plains grasslands. Soil Sci. Soc. Am. J. 51:1173–1179.

Parton, W.J., J.W.B. Stewart, and C.V. Cole. 1988. Dynamics of C, N, P, and S in grassland soils: A model. Biogeochemistry 5:109–131.

Parton, W.J., J.M.O. Scurlock, D.S. Ojima, T.G. Gillmanov, R.J. Scholes, D.S. Schimel, T. Kirchner, J.C. Menaut, T. Seastedt, E. Garcia Moya, A. Kamnalrut, and J.L. Kinyamario. 1993. Observations and modeling of biomass and soil organic matter dynamics for the grassland biome worldwide. Global Biochem. Cycles 7:785–809.

Parton, W.J., and P.E. Rasmussen. 1994. Long-term effects of crop management in a wheat-fallow system: II. CENTURY model simulations. Soil Sci. Soc. Am. J. 58:530–536.

Paustian, K., W.J. Parton, and J. Persson. 1992. Modeling soil organic matter in organic-amended and nitrogen-fertilized long-term plots. Soil Sci. Soc. Am. J. 56:476–488.

Ryan, M.G., J.M. Melillo, and A. Ricca. 1990. A comparison of methods for determining proximate carbon fractions of forest litter. Can. J. For. Res. 20:166–171.

Shaffer, M.J. 1985. Simulation model for soil erosion-productivity relationships. J. Environ. Qual. 14:111–150.

Sorensen, L.H. 1981. Carbon-nitrogen relationships during the humification of cellulose in soils containing different amounts of clay. Soil Biol. Biochem. 13:313–321.

Tate, K.R., and B.K.G. Theng. 1980. Organic matter and its interactions to inorganic soil constituents. p. 225–249. *In* Soils with variable charge. N.Z. Soc. Soil Sci., Lower Hutt, NZ.

Theng, B.K.G. (ed.). 1980. Soils with variable charge. N.Z. Soc. Soil Sci., Lower Hutt, NZ.

Thurchenek, L.W., and J.M. Oades. 1979. Fractionation of organo-mineral complexes by sedimentation and density techniques. Geoderma 21:311–343.

van Veen, J.A., and E.A. Paul. 1981. Organic C dynamics in grassland soils. I. Background information and computer simulation. Can. J. Soil Sci. 61:185–201.

van Veen, J.A., J.N. Ladd, and M. Amato. 1985. Turnover of carbon and nitrogen through the microbial biomass in a sandy loam and a clay soil incubated with [^{14}C(U)] glucose and [^{15}N](NH$_4$)$_2$SO$_4$ under different moisture regimes. Soil Biol. Biochem. 17:747–756.

van Veen, J.A., J.N. Ladd, and M.J. Frissel. 1984. Modelling C and N turnover through the microbial biomass in soil. Plant Soil 76:257–274.

Vitousek, P.M., D.R. Turner, W.J. Parton, and R.L. Sanford, Jr. Litter decomposition on the Mauna Loa environmental matrix, Hawaii: Patterns, mechanisms, and models. Ecology (In press.)

Williams, J.R., J.W. Putman, and P.T. Dyke. 1985. Assessing the effect of soil erosion on productivity with EPIC. In: Proc. National Symp. Erosion and Soil Productivity. New Orleans, LA, December 1984. American Society of Agricultural Engineers. St. Joseph, MI.

10 A Spatial Framework for Integrating Soil–Landscape and Pedogenic Models

B.K. Slater, K. McSweeney, S.J. Ventura, and B.J. Irvin

University of Wisconsin
Madison, Wisconsin

Alex B. McBratney

University of Sydney
Sidney, Australia

Models of pedosphere processes have increasing utility for understanding the fundamental dynamics of soil systems and in managing the environment and forecasting environmental change. Developments in modeling evolutionary changes within the soil system (as evidenced by papers in this volume) accompany models of external processes and linkages between soil and other components of the biosphere, hydrosphere, and geosphere. An integration of models describing processes continuous between the soil system and the surrounding environment, linking pedological, earth surface and ecological processes, is timely. Such integration is essential to an understanding of complex historical processes such as soil genesis, and in prediction and management of current and future environmental change at global, regional, and local scales.

Pedogenic process models need integration with physical models of real soil landscapes. Hoosbeek and Bryant (1992) review pedogenic modeling and the quantification of pedogenic processes. While pedogenic models may adeptly describe dynamic soil processes, they are often applied to abstract situations rather than being linked to a realistic space–time context. Burrough (1989) discusses the lack of consideration of spatial variability of control parameters and data in many models. Conditional representations of soil distribution in time and space are needed to provide detailed information for models at all points in three-dimensional (3-D) space and time, rather than generalized data for static areas at a single scale.

Soil–landscape models are expressions of the linkage of geosphere and biosphere components as products of historical process. Consequently, they establish predictive relationships between components and offer a real spatial context for application of pedogenic models. Opportunities for integration of realistic soil information and process models have been enhanced by advances in soil data acquisition and processing, information systems for data storage and analysis, and visualization techniques.

The integration of soil distribution data and pedogenic models may be useful at a variety of scales, but recent advances in development of soil–landscape models provide particular opportunities for applying models at the landscape or catena scale. Terrain analysis has shown particular promise for modeling surface hydrological processes that are important concepts in pedogenesis. Geomorphometric terrain attributes have been derived from digital terrain models to provide representations of the configuration of landscape surfaces (Moore et al., 1991). The addition of soil and regolith stratigraphic components, derived by quantitative methods, will permit a better representation of the soil–landscape in 3-D (McSweeney et al., 1994). This chapter proposes a framework for analysis and representation of soil–landscapes to provide an appropriate interface with component process models of soil evolution. Major components of the framework are shown in Fig. 10–1.

A FRAMEWORK FOR LINKING PEDOGENIC AND LANDSCAPE MODELS

Approach

A suitable framework for linking soil–landscape and pedogenic models involves development of a conditional, mechanistic, and stochastic representation of soil diversity. To provide a spatial context for pedogenic modeling, the representation needs to be conditioned on reality. The model must incorporate processes at appropriate time–space scales. A stochastic approach is needed because of uncertainty in inputs and processes.

The initial outcome of the approach is a quantitative, 3-D description of the soil landscape at a given fixed point in time. Evolution of the soil landscape may be represented by animating a sequence of model constructions for sequential points in time. It is envisaged that the spatial model will serve as an initial state for dynamic modeling of processes in the soil landscape such as that proposed by Kirkby (1985).

Three general principles govern the conditional model:

1. Explicitly accommodate the patterns and processes that result in soil diversity.
2. Incorporate information on the continuously varying nature of real soil landscapes.
3. Develop data structures that provide for flexible integration with component process models of soil genesis.

SOIL DIVERSITY—PATTERN AND PROCESS

Spatial and Temporal Scale

The issue of scale is important in considering the integration of soil data and models (Kachanoski, 1988). The discussion of spatial and temporal scales and the

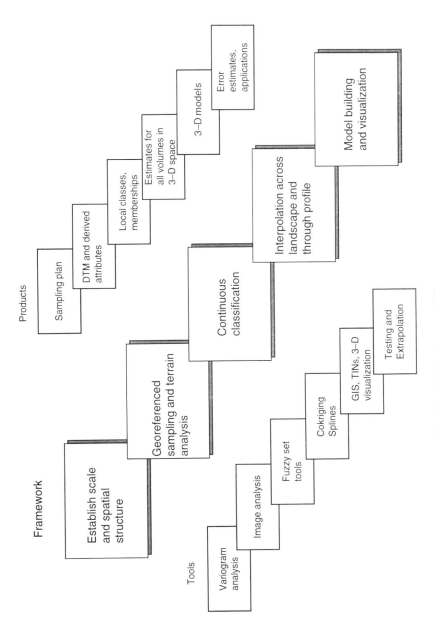

Fig. 10-1. Framework for a digital spatial model of soil-landscape variability.

postulated hierarchical nature of scale relationships in nature has been a recent preoccupation of geomorphic and ecological literature. Hierarchy theory (Allen & Starr, 1982) formalizes the view that natural processes operate at various scales and frequencies in time and space, and are expressed as structural and functional patterns. These patterns of material and process are the basis of soil diversity. The pedosphere can be viewed as a heterogeneous continuum, and pattern and diversity are expressions of the varying degrees of discontinuity and disturbance.

Haigh (1987) analyzes the interactions between scales at various levels in the landscape. Dijkerman (1974), Haigh (1987), and Hoosbeek and Bryant (1992) present holarchies (nested hierarchical structures) which attempt to formalize levels of organized structure that exist in the soil system. The levels vary from the molecular to global scale. McSweeney et al. (1994) present another scheme which illustrates the continuum of scales which exist in the geosphere. Similar schemes for ordering of relief unit sizes are presented by Dikau (1989). Soil genetic processes operate within recognizable scales.

Examining Scale

Models need to be linked to a particular spatial and temporal scale. Since linkages between pattern and process can occur across scales, there is a need to understand how processes at finer scales integrate. Nonlinear relations between outputs and inputs create particular problems.

Soil scientists often consider scale in terms of cartographic resolution— the relationship between a geographic measurement and its representation on a map. In the cartographic context, large scale refers to a detailed map resolution. In considering the general issue of scale as a spatial or temporal dimension, a more consistent terminology is helpful. Turner and Gardner (1991) provide a series of definitions of scale-related phenomena. In this context, *fine scale* applies to higher precision of measurement or higher resolution and to a smaller dimension of the phenomenon of study; while *broad scale* refers to coarser resolution and larger dimension of the object of interest. Landscape ecologists have developed some principles for examining scale linkages (Risser, 1987). From a given scale of interest, the next broader level serves as a boundary constraint and may be used to gauge the significance of the scale of interest. Phenomena observed at finer scales may explain processes that exert control at the broader scale (O'Neill, 1987). As the broadness of the scale of interest increases, the significant constraining variables become fewer, more apparent (emergent) and/or dominant (Meentemeyer & Box, 1987). For example, at the continental scale, climate may become the most important constraint on an attribute such as soil pH, but which is locally controlled by many other influences at finer scales (Folkoff et al., 1981). Early qualitative pedogenic models and national soil classification schemes formalized these large-scale influences [e.g., the emphasis on pedocals and pedalfers (Jenny, 1941) and the zonal concept].

Scale relationships are rarely linear and patterns rarely defined uniformly across scales, so simple additive models are likely to be inadequate to extrapolate soil–landscape structure and function between scales (Risser, 1987). The use of common indices that do not change fundamentally across scales (i.e., which vary only in grain or degree of observational detail) is one useful approach. Another is the development of linking models that integrate component models for each scale (Risser, 1987). Fractal models offer a means of structuring scale, but self-similarity implies that attributes are not uniquely influenced at particular scales.

Unlike temporal scales, spatial scales may vary depending on direction or orientation. The scale and spatial structure of processes and attributes differs vertically down the profile and across the landscape. Soil organic C content typically exhibits distinct surface and profile distribution patterns that can be largely explained by processed influencing C dynamics (Fig. 10–2). Complex hydrologic responses may be conditioned by local juxtapositions of contrasting soil materials and macropores that have an isolated distribution or which vary greatly over short distances. Generalized attribute values applied at too broad a scale may miss influential finer-scale phenomena.

In establishing relationships, scale needs to be examined for all significant attributes, and observation density matched to the innate spatial scale and resolution requirements of the model. The structure and function—of the phenomenon of interest (diversity, degree of discontinuity or variability)—should determine the scale of measurement.

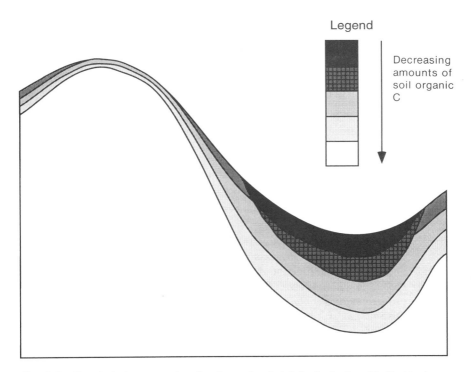

Fig. 10–2.   Hypothetical representation of surface and vertical C distribution in an idealized landscape.

Land Surface Modeling—Geomorphic Attributes
and Digital Terrain Models

Quantitative tools for examining land surface configuration are available to rapidly derive salient landform attributes. These tools are valuable for modeling soil landscapes and for defining relationships between soil and landform parameters. Foremost among the tools is digital image processing, which is used to derive secondary landform attributes from digitized elevation datasets. The early work on geomorphometric image processing by Evans (1980) and Zevenbergen and Thorne (1987) has recently been applied to studies of soil landscapes (Moore et al., 1993). A range of landform attributes including slope, aspect, profile and plan curvature and catchment areas can be derived by the application of simple algorithms and moving window techniques to digital terrain models (Moore et al., 1993, McSweeney et al., 1994).

Subsurface Modeling—Soil Data Sources

Component process models, whether simulating pedogenic evolution or other dynamic changes in biosphere components, require data on the spatial and temporal distribution of soil attributes in soil landscapes. Traditional sources for this information, such as soil surveys, and data structures for manipulating the information may not be appropriate for the needs of models.

Soil surveys are the most comprehensive and commonly used geographic data available on soil materials for model input. Site data directly gathered in soil survey may serve as model input, or data may be generalized for geographic areas or map units. In other cases, interpretations of accessory properties derived from surveys also have been applied. Soil surveys have necessarily involved standardized methods for data collection. The methods have determined to a large extent the quality and quantity of the data, and the form in which it is represented to users. An understanding of the limitations of soil survey is necessary if the data are to be appropriately used as model input. Table 10–1 contrasts traditional soil data sources with the alternative approach discussed here.

The challenge of providing appropriate soil data for models can be met by accepting a more flexible series of techniques and data structures for the representation of soil spatial and temporal variability. The conventional soil survey paradigm (Hudson, 1992) can be extended to encompass a more generalized and integrated series of hypotheses relating soil spatial features with other biospheric components:

1. Domains exist within the landscape where soil, hydrologic, and landform attributes are products of common genesis and where soil, water and land surface function in an integrated manner.
2. Domains within broader landscape divisions show coherent spatial patterns related to *common, coupled* and *restricted* events and processes.
3. Distinct landscape divisions exhibit disjunct patterns of soil diversity and few apparent threads of common genesis, yet have coevolved in response to transcending events and processes.
4. Soil–landscape relationships developed in a local area can be extrapolated to areas that evolved under similar geological and environmental conditions.

Table 10–1. Comparison of traditional soil data sources and those required for proposed three-dimensional (3-D) soil–landscape model.

Features	Traditional soil survey	Proposed model
Purpose	Data gathered for variety of purposes	Data gathered specifically for model
Classification	National and regional classification schemes	Local classification from landscape
Scale	Standard scales applied universally	Scale appropriate to local diversity and model need
Variability	Estimates not usually included	Measured locally
Sample intensity	May not reflect local diversity, often constant for a given survey	Determined by local diversity, variable across landscape
Spatial delineations	Discrete, discontinuous and closed	Continuous
Taxonomic units	Rigid defined boundaries and limits	Continuous membership in multiple classes
Point data	Site data collected to define mapping units	Data from georeferenced sites
Boundary observation	Most frequently inferred from other distributions (vegetation, landform) seen on aerial photographs	Interpolated from site data
Attributes	Derived from classified map units, change at unit boundaries	Variance structure may be unique for each attribute, continuous variation
Modal profiles	Apply to all occurrences of unit	Statistically derived from sample data
Accessory properties	Correlations not always examined	Covariance established locally
Dimensions	Data generalized to two dimensions (2-D)	Three-dimensional (3-D) model and data structures
Model Type	Deterministic, empirical	Stochastic, mechanistic

An idealized representation of a silt mantled, bedrock cored soil–landscape from the Driftless Area of southwestern Wisconsin illustrates these relationships (Fig. 10–3). Domain A soils developed in several layers of transported silt, and now form a sloping terrace. Soils of the terrace exhibit fairly uniform morphologic features. Attributes including horizon thickness, color, texture, and subsoil hydromorphic features can be explained by *common* pedogeomorphic processes that are not exclusively shared by adjacent soils. The soils of Domain B are formed in a variety of contrasting materials that accumulated as layers during multiple erosional, depositional, and mass wasting events on the hill slope. Attributes such as depth to bedrock and horizon thickness are highly variable. The result is a diverse

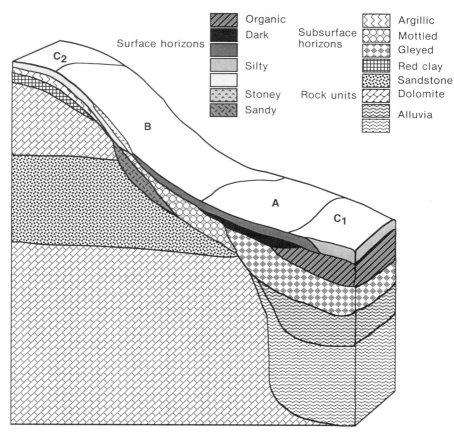

Fig. 10–3. Schematic cross-section of a segment of Driftless Area landscape (southwestern Wisconsin) showing major domains and horizons.

soil pattern that exhibits threads of common origin, but that is locally differentiated by the nature and arrangement of layers and hydrologic processes. Domains C_1 and C_2 contain strongly contrasting soil. The source of all the silt mantle overlying soils in domain C_1 is related to erosion of materials at higher elevations in the subwatershed, including ridgetops (Domain C_2). Distinct stratigraphic markers, which include thin bands of transported red "residuum" and coarse fragments, mark *coupled* erosional and depositional events. Layers eroded from one domain may be deposited in another in reverse stratigraphic order.

CONTINUOUSLY VARYING NATURE OF THE SOIL LANDSCAPE

Soil Variability

Geostatistical tools are now widely used to define the source, scale, and nature of soil variation. A general approach to collecting soil data appropriate for models is analogous to the scientific method—gather information, propose hypotheses for underlying structure of the variation observed, and test elsewhere to measure efficacy of models.

A systematic approach thus involves the following steps:

1. Reconnaissance to measure scale of variability and to generate variograms for salient attributes.
2. Design of sampling schemes to measure attributes at appropriate density and scale.
3. Interpolation to generate a spatial model of attribute distribution at all points.
4. Testing by sampling at "unknown" points.

A more comprehensive approach to gathering soil data includes use of traditional methods. Diversity measurements can be incorporated into traditional mapping units where information is available or specifically collected.

Traditional survey information and other prior data also can be incorporated. It is prudent to critically utilize a priori knowledge as a guide to sampling design. For example, the landscape can be stratified according to known geographic divisions (including soil map units), and sampling strategies may be varied according to known environmental gradients such as slope attributes. Assumptions made in analytical techniques, such as stationarity in geostatistical models, need to be examined locally. Digital processing of terrain models can be applied prior to field survey to assist sampling design. Local classes with defined ranges of landform attributes including slope, aspect, curvature and catchment area can be derived as an alternative to arbitrary divisions of slope.

Representing Soil Continuity

Soil scientists often refer to the pedosphere as a continuum (McBratney et al., 1992). Soil classification and surveys and their end-product, the soil map, have institutionalized the alternative view that discretely bounded soil individuals exist across the landscape. The soil survey paradigm (Hudson, 1992) of map unit delineations locating occurrences of rapid change in one or more soil forming factors is understood intuitively by experienced field soil scientists, but not necessarily by users of soil surveys, including builders and operators of models. Discontinuities in the geographic distribution of soil attributes are common, but they may in fact be disjoint; the presence of *connected* boundaries between discrete soil bodies can be misleading. Soil boundaries have variable width (Fridland, 1976) and the degree of distinctness varies along the boundary. While some soil properties may change rapidly where discontinuities of the soil forming factors are dominant or coincident, other properties change gradually or at displaced locations (Butler, 1964). Boundaries of individual attributes within layers or of layers themselves are not necessarily spatially coincident.

Distinct boundaries also are institutionalized in traditional descriptions of the 3-D of the soil profile. Boundaries between soil horizons or layers may be very gradual and the various soil properties may change at different rates and exhibit complex relationships and disparate scales. Soil classification schemes based on rigid classes are often based on identification of a limited number of diagnostic horizons and may ignore the complexity of subtle layering and continuous change in the vertical direction. Such complexity is especially critical where compound horizons are formed by bioturbation and other soil mixing processes (e.g., A/B horizons in mollic intergrades) or where complex horizons are formed by temporally separated regimes of pedogenesis (e.g., calcified argillic horizons, welded paleosols).

The traditional abruptly bounded or discontinuous soil unit can be viewed as subclass or end-member of a series of continuous possibilities. Developments in continuous classification and mapping show promise for ordering soil spatial data for models. Continuous classes may apply to both character (attribute) and geographic domains (space/time) and hence are appropriate for both classification and mapping.

Recently continuous classification methods have been applied to soil profiles and to soil horizons or layers (Odeh et al., 1990; Powell et al., 1991; McBratney et al., 1992; McBratney & De Gruitjer, 1992). Continuous classification methods utilize fuzzy set concepts. An individual is permitted simultaneous membership in the universe of classes defined for the data set. Classes may be derived by numerical methods, and individuals allocated class membership using simple statistical measures of distance in attribute space (Ward et al., 1992). Hence, each individual (whether profile, layer, mapping unit, or arbitrary volume unit) can be described by a vector of memberships in the series of classes. The membership indicates how similar the individual is to the centroids of each class. Individuals with high membership in only one class are very similar to the class centroid, while those with a significant degree of membership in several classes are intermediates in the continuum. The class memberships can be used for mapping or visualizing the spatial distribution of the individuals, and the membership vectors are particularly suited to geostatistical or other interpolation techniques.

Horizons, Layers and Other Soil Entities

Model input data may need to be generalized or classified into flexibly defined entities or soil units, though in many instances individual data values and point data sets may be more useful than classified data. Models of landscapes should be derived locally, with entities then related to broader schemes such as national soil classifications (Butler, 1980).

Many studies compare soil profiles sampled along a transect to larger soil landscape units, which may result in a conflict of scale (Gerrard, 1990). The pedon concept assumes that the most important direction of water movement is vertical and does not explicitly accommodate lateral processes such as movement of water and its constituents. A better perspective may be gained by analyzing representative soil characteristics within each landform element rather than using a pedon or two to represent the soils of an area.

Entities include traditional whole soil individuals such as pedons, polypedons and map units, or more discrete units such as horizons or layers. Points themselves may considered the fundamental entities in space and time (Holmgren, 1988), though attributes are practically measured over areas and volumes. For 3-D models, it may be appropriate to locally or universally define appropriate volume units. These may have variable or flexible size and dimensions, be capable of hierarchical organization, and be aggregated by generalization rules derived from studies of scale. Further advantages for modeling accrue from the accommodation of spatial and temporal feedbacks.

For some modeling purposes, the soil horizon may be the appropriate individual. Horizons are more discrete and internally less variable than traditional whole-soil entities. Horizons or soil strata can be defined locally without constraints of established taxonomies. In some cases, horizons are entities with a common genetic context. In other cases, complex and compound horizons result from interaction of a variety of processes and material sources. Since soil is continuous with other earth surface materials, a "stratigraphic" approach focusing on individual soil layers can assist in integration of soil process models with models from other disciplines. A stratigraphic approach also is appropriate for the many soil landscapes where transported materials are superimposed, and where overlying soil horizons do not vary exactly in tandem. Horizons are conceptually appropriate for generalizing functional attributes such as hydrology and can be readily modeled as volume entities (Bouma, 1989).

Horizons and layers do not resolve the difficulty of spatial continuity of soil attributes and the varying rates of change of many soil attributes down the profile. For modeling purposes, arbitrary volume units may be more appropriate carriers of attribute information.

DATA STRUCTURES FOR FLEXIBLE INTEGRATION WITH COMPONENT PROCESS MODELS

Geographic information systems (GIS) and scientific visualization have provided many options for the management of soil data needed for models. Advantages and limitations of available and projected data structures need to be assessed so that appropriate choices are made for each modeling purpose (Jones, 1989).

Representation of Soil Mapping Units in Two-Dimensional Vector-Based Geographic Information Systems

Vector-based GIS software packages are commonly used for soil applications. Screen displays closely resemble paper maps. Consequently, problems similar to those associated with two-dimensional (2-D) maps present limitations for process modeling applications. For example, 3-D phenomena are depicted in 2-D with no depth. Digital soil maps can be easily linked with data base management systems and readily integrated with other geographic data such as land ownership, hydrology, and slope. Multiple attributes of mapping units can be readily displayed and no limits are imposed on the degree of precision, because map unit polygons can be of any size.

These systems recognize that spatial relationships exist between map features. This explicit adjacency or connectivity among polygons, lines and points (topology) allows for the modeling of relationships between features. Vector-based systems are used for "weighting and rating" models (groundwater contamination susceptibility), for aspatial surface processes (litter decomposition), and to provide attributes for other models (productivity, infiltration).

Transitions between soil types are represented by bounded finite delineations. Homogeneous units inadequately represent mixtures of soil types. For this reason, vector-based systems do not easily support geostatistical and fuzzy set tools. It also is difficult to link discontinuous 2-D polygons with dynamic models of landscape processes.

Representation of Soils and Surfaces in Two and One-Half Dimensional Raster-Based Geographic Information Systems

Raster systems store information in grid cells which are generally of fixed size. Raster software packages are in widespread use allowing for easy integration with other geographic data. This structure lends itself to model applications, such as the prediction of erosion rates or wildlife habitat suitability, as well as to "finite element" surface process modeling, such as flow routing. Precision in the x and y directions is limited by cell size. If the cell size is 30 m^2, then soil units must be in increments of 30 m^2. Trade-offs exist between cell size (resolution) and data volume. The issue of whether average data for each cell is used or whether the cell is represented by point or other data needs to be addressed. Also, only one attribute value is allowed per cell, so multiple attributes require multiple files. The cell structure facilitates the use of geostatistics though explicit topology is not present. Some relief from the restrictions of uniform cell size is afforded by the use of more complex data structures such as octrees and quadtrees, but at a cost of high computational intensity.

Digital terrain models store elevation data in the cells that can be used to create pseudo 3-D displays such as wire net, terrain drapes, and shaded relief. A wire net consists of nodes and connecting lines. Digital terrain models can be readily and accurately derived using photogrammetric methods and commercial software packages. Derivable landscape attributes include slope, aspect, shape, catchment basins, ridgelines and drainageways. Multiattribute landform units also may be derived by image processing. Advances in derivation of geomorphic units from digital terrain models show promise for modeling soil–geomorphic relationships.

Representation of Soil Volumes in Three-Dimensional Vector-Based Data Structure (Solids Modeling)

Solids modeling, commonly seen in drafting and engineering design applications, is an algorithmically complex vector-based structure. This form of modeling provides visualization of shape only with no topology. If they existed, true holographic 3-D displays could provide a very accurate representation of stratigraphy and topologic relations. Precision would be limited only by the sampling density, though the continuous nature of the representation may imply a greater precision than is justified. At this time, there is no commercial software with the features necessary for modeling soil landscapes in 3-D; limited prototype software has potential.

Representation of Soil Stratigraphic Surfaces in Two and One-Half-Dimensional Triangulated Irregular Network Data Structure

A triangulated irregular network (TIN) is a set of adjacent triangles developed from irregularly spaced points having x, y, and z values. Topological relationships between points and triangles allow for generation of surface models, including pseudo 3-D displays (wire nets and profiles). An explicit relationship exists between sample points (nodes) and the stratigraphic surface. Elevations are accurate at the nodes but are interpolated between nodes. Therefore, positional accuracy is tied to the TIN nodes, not the surface configuration. Commercial TIN software is designed for modeling of single surfaces, extensions are needed for linked multiple surfaces such as superimposed horizon boundaries. The TIN surfaces can be used to derive slope gradient and aspect for landscapes, estimate deposition or erosion at sample points, as well as model other temporal processes. The TIN systems may be less adept at handling continuous attribute distributions than modeling abrupt boundaries and surfaces. Figure 10–4 shows three idealized TIN models representing interfaces between soil horizons. The TIN surfaces were derived by interpolation from a 5 by 10 grid of x, y coordinates for surface elevations and depth (z) to two subsoil horizons.

Representation of Soil Volumes in Three-Dimensional Volume Cell (Voxel) Data Structure

Voxels can be thought of as stacked boxes each occupying a fixed volume and containing certain attributes, or as 3-D versions of raster pixels. This structure allows for pseudo 3-D displays of profiles or level slices across a constant depth. Vowels provide an explicit relationship between surface area and volume; however, cell size limits the $x, y,$ and z precision of voxels. This approach is reasonably simple algorithmically, but no explicit topology is present. Commercial software using voxels is not readily available, but the structure may possibly be derived from raster-based systems. Geological strata and structures have recently been modeled using 3-D grid methods (Hamilton & Jones, 1992). Development of this structure could have positive implications for finite element-based models such as those used to simulate groundwater movement or water table configurations.

The voxel structure can be flexibly applied at any data scale and may be useful for modeling soil systems from thin sections to the entire pedosphere. In order to represent short-distance variability, a large number of small voxels are needed with consequent high-data storage requirements. Encoding techniques such as octrees (Samet, 1990), which permit variable size volume units could be used to match attribute variation with cell size. For example, homoscedastic voxels, which are volume units encapsulating constant attribute variance in spatial or temporal dimensions, and hence of varying size, could be constructed. These voxels are similar in concept to vector-based solid models.

Figure 10–5 depicts a soil–landscape or catena segment where each voxel represents a located volume of soil with a series of measured or derived attributes, and is surrounded either by other voxels with similar or contrasting attributes or by nonsoil material. Voxels represent a defined space through which energy and matter flow can be simulated. The voxel approach is particularly suitable for representing soil attributes for 3-D finite element modeling and flow routing. Rates and direction of flow and partitioning of matter and energy may be derived from attributes of voxels and from combinations of adjacent voxels and pathway characteristics.

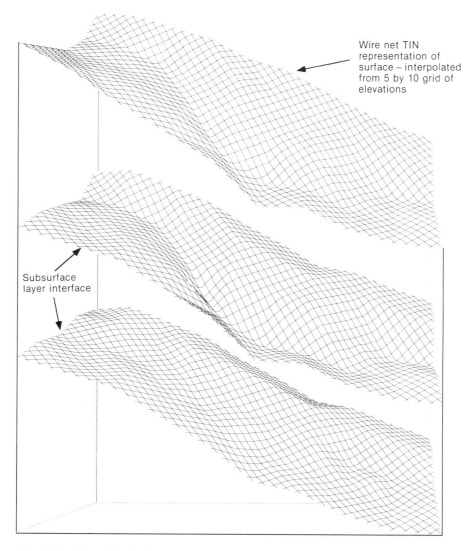

Wire net TIN
representation of
surface – interpolated
from 5 by 10 grid of
elevations

Subsurface
layer interface

Fig. 10–4.   Hypothetical schematic triangulated irregular network (TIN) representation of soil surface and subsurface horizon interfaces.

SUMMARY

The provision of a real spatial context for the application of dynamic pedogenic models is a major challenge. Realistic spatial models of soil landscapes
require the collection of large amounts of data at a scale appropriate to the needs
of users, but responsive to the spatial scale and variability of the soil–landscape.
An inflexible application of survey or sampling techniques is unlikely to yield

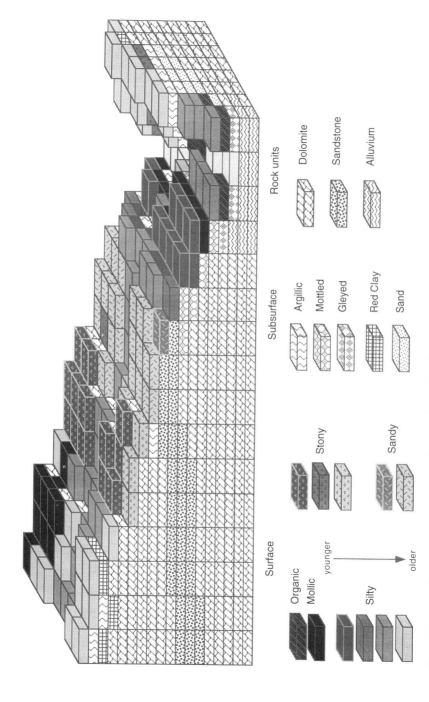

Fig. 10–5. Simplified voxel representation of surface materials and landform of a portion of the Driftless Area, southwestern Wisconsin.

high-quality information needed for modeling pedogenic processes in time and space. We maintain that explicit accommodation of the structure of scale and diversity of landscapes is needed. A data model conditioned on reality demands recognition of the 3-D and continuous nature of soil attribute distribution.

The framework we propose generates a conditional, mechanistic, stochastic model of soil attribute distribution in space, and serves as a snapshot of the soil landscape at a particular point in time. The model is developed by scale-sensitive collection of soil and landform attributes in the field. The landform surface is modeled by application of digital processing techniques to elevation data sets. Subsurface components are integrated by 3-D interpolation of attribute data from georeferenced sample points. Data storage and visualization of the information can be achieved by use of appropriate 3-D data structures. Voxel-based data structures will permit storage and visualization of soil attribute data generated for all blocks of soil in a soil landscape. By changing the attributes and even the presence or absence of volume units, the dynamics of soil-landscape evolution can be examined and visualized. The approach is particularly relevant to pedogenic models, which simulate matter and energy partitioning.

ACKNOWLEDGMENTS

The work reported in this manuscript was supported in part by a grant from the NSF (SES-0210093).

REFERENCES

Allen, T.F.H., and T.B. Starr. 1982. Hierarchy: Perspectives for ecological complexity. Univ. Chicago Press, Chicago.

Bouma, J. 1989. Using soil survey data for quantitative land evaluation. p. 225–239. *In* B.A. Stewart (ed.) Advances in soil science. Vol. 9. Springer-Verlag, New York.

Burrough, P.A. 1989. Matching spatial databases and quantitative models in land resource assessment. Soil Use Manage. 5:3–8.

Butler, B.E. 1964. Can pedology be rationalized? A review of the general study of soils. Australian Soc. Soil Sci. Publ. 3. Canberra, Australia.

Butler, B.E. 1980. Soil classification for soil survey. Clarendon Press, Oxford, England.

Dijkerman, J.C. 1974. Pedology as a science: The role of data, models and theories in the study of natural systems. Geoderma 11:73–93.

Dikau, R. 1989. The application of a digital relief model to landform analysis in geomorphology. p. 51–77. *In* J. Raper (ed.) Three dimensional applications in geographical information systems. Taylor and Francis, Ltd., New York.

Evans, I.S. 1980. An integrated system of terrain analysis and slope mapping. Z. Geomorphol. N.F. Bodenkd. 36:274–295.

Folkoff, M.E., V. Meentenmeyer, and E.O. Box. 1981. Climatic control of soil acidity. Phys. Geogr. 2:116–124.

Fridland, V.M. (ed.). 1976. Soil combinations and their genesis. USDA-ARS, N. Amerind Publ. Co., New Delhi, India.

Gerrard, A.J. 1990. Soil variations on hillslopes in humid temperate climates. Geomorphology 3:225–244.

Haigh, M.J. 1987. The holon: Hierarchy theory and landscape research. p. 181–192. *In* F. Ahnert (ed.) Geomorphological models—Theoretical and empirical aspects. Braunschweig, Catena Suppl. 10. Elsevier Publ., Amsterdam.

Hamilton, D.E., and T.A. Jones (ed.). 1992. Computer modeling of geologic surfaces and volumes. Am. Assoc. Petrol. Geol., Tulsa, OK.

Holmgren, G.G.S. 1988. The point representation of soil. Soil Sci. Soc. Am. J. 52:712–716.

Hoosbeek, M.R., and R.B. Bryant. 1992. Towards the quantitative modeling of pedogenesis—a review. Geoderma 55:183–210.

Hudson, B.D. 1992. The soil survey as a paradigm-based science. Soil Sci. Soc. Am. J. 56:836–841.

Jenny, H. 1941. Factors of soil formation: A system of quantitative pedology. McGraw-Hill, New York.

Jones, C.B. 1989. Data structures for three-dimensional spatial information system in geology. Int. J. Geogr. Inform. Sys. 3:15–31.

Kachanoski, R.G. 1988. Processes in soils—from pedon to landscape, p. 153–177. In T. Rosswall et al. (ed.) Scales and global change. John Wiley & Sons, Chichester, England.

Kirkby, M.J. 1985. A basis for soil profile modelling in a geomorphic context. J. Soil Sci. 36:97–121.

McBratney, A.B., J.J. De Gruitjer, and D.J. Brus. 1992. Spacial prediction and mapping of continuous soil classes. Geoderma 54:39–64.

McBratney, A.B., and J.J. De Gruitjer. 1992. A continuum approach to soil classification by modified fuzzy k-means with extragrades. J. Soil Sci. 43:159–175.

McSweeney, K., P.E. Gessler, B.K. Slater, R.D. Hammer, J. Bell, and G.W. Petersen. 1994. Towards a new framework for modeling the soil-landscape continuum. p. 127–145. In R. Amundson et al. (ed.) Factors of soil formation: A fiftieth anniversary retrospective. SSSA Spec. Publ. 34. SSSA, Madison, WI.

Meentemeyer, V., and E.O. Box. 1987. Scale effects in landscape studies. p. 15–34. In M.G. Turner (ed.) Landscape heterogeneity and disturbance. Springer Verlag, New York.

Moore, I.D., P.E. Gessler, G.A. Nielsen, and G.A. Peterson. 1993. Soil attribute prediction using terrain analysis. Soil Sci. Soc. Am. J. 57:443–452.

Moore, I.D., R.B. Grayson, and A.R. Ladson. 1991. Digital terrain modelling: A review of hydrological, geomorphological, and biological applications. Hydrol. Proc. 5:3–30.

O'Neill, R.V. 1987. Hierarchy theory and global change. p. 29–45. In T. Rosswall et al. (ed.) Scales and global change. John Wiley & Sons, Chichester, England.

Odeh, I.O.A., A.B. McBratney, and D.J. Chittleborough. 1990. Design of optimal sample spacings for mapping soil using fuzzy k-means and regionalized variable theory. Geoderma 47:93–122.

Powell, B., A.B. McBratney, and D.A. McCleod. 1991. The application of fuzzy classification to soil pH profiles in the Lockyer Valley, Queensland, Australia. Catena 18:409–420.

Risser, P.G. 1987. Landscape ecology: state of the art. p. 4–14. In M.G. Turner (ed.) Landscape heterogeneity and disturbance. Springer Verlag, New York.

Turner, M.G., and Gardner, R.H. 1991. Quantitative methods in landscape ecology: An introduction. p. 4–14. In M.G. Turner and R.H. Gardner (ed.) Quantitative methods in landscape ecology: The analysis and interpretation of landscape heterogeneity. Springer Verlag, New York.

Samet, H. 1990. Applications of spatial data structures: Computer graphics, image processing, and GIS. Addison Wesley, Reading, MA.

Ward, A.W., W.T. Ward, A.B. McBratney, and J.J. De Gruijter (ed.). 1992. MacFUZZY—A program for data analysis by generalized fuzzy K-means on the Macintosh. CSIRO Div. of Soils., Div. Rep. 116, Canberra, Australia.

Zevenbergen, L.W., and C.R. Thorne. 1987. Quantitative analysis of land surface topography. Earth Surf. Processes and Landforms 12:47–56.